William Hooper, Jacob F. Bielfeld

The Elements of Universal Erudition

containing an analytical abridgment of the sciences, polite arts, and belles lettres -

Vol. 3

William Hooper, Jacob F. Bielfeld

The Elements of Universal Erudition
containing an analytical abridgment of the sciences, polite arts, and belles lettres - Vol. 3

ISBN/EAN: 9783337394882

Printed in Europe, USA, Canada, Australia, Japan

Cover: Foto ©berggeist007 / pixelio.de

More available books at **www.hansebooks.com**

THE
ELEMENTS
OF
UNIVERSAL ERUDITION,

CONTAINING AN

ANALYTICAL ABRIDGMENT

OF THE

SCIENCES, POLITE ARTS,

AND

BELLES LETTRES,

By Baron BIELFELD,

Secretary of Legation to the King of Prussia, Preceptor to Prince Ferdinand, and Chancellor of all the Universities in the Dominions of his Prussian Majesty, Author of the Political Institutes, &c.

Indocti discant, & ament meminisse periti.

TRANSLATED FROM THE LAST EDITION PRINTED
AT BERLIN

By W. HOOPER, M.D.

VOL. III.

LONDON:
Printed by G. Scott,
For J. ROBSON Bookseller in New-Bond Street,
and B. LAW in Ave-Mary Lane.
MDCCLXX.

CONTENTS

OF THE

THIRD VOLUME.

Book the Third.

Chap.	Page
I. Of the Belles Lettres, and the sciences of the memory in general	1
II. Of mythology	15
III. Chronology	36
IV. Of history in general, and of its divisions	69
V. Ancient history	90
VI. The history of the middle Age	123
VII. Modern history	145
VIII. Ecclesiastic history	187
IX. The history of the Christian Church, of heresies, popes and reformers	207
X. Of antiquities	231

CONTENTS.

Chap.		Page
XI. Medals and coins		244
XII. Diplomatics		260
XIII. Statistics		268
XIV. Of travels and travellers		279
XV. Of Geography		287
XVI. Genealogy		303
XVII. Blazonry		310
XVIII. Of philology in general		319
XIX. Of the oriental languages		328
XX. Of the other learned languages, and of paleography		337
XXI. Modern languages		349
XXII. Digression on exercises		359
XXIII. ———— on certain anomalous arts and sciences		377
XXIV. ———— on chimerical arts and sciences		385
XXV. ———— on schools, colleges, universities and academies		402
XXVI. The history of the sciences		414
XXVII. Of the knowledge of authors, and of biography		427
XXVIII. Digressions, 1. On criticism; 2. On literary journals; 3. On libraries		437

BOOK THE THIRD.

CHAP. I.

OF THE

BELLES LETTRES,

AND THE

Sciences of Memory in General.

I. WHETHER we consult the voluminous dictionaries of the French language, or those treatises that profess to point out the method of studying and teaching the Belles Lettres, we find not, in the one or the other, either a clear definition, or a succinct explication of the words Belles Lettres, nor any summary of those sciences which are comprehended under that general and collective denomination. It appears to be a vague term, under which every one may include what-ever

ever he thinks proper. Sometimes we are told that by the Belles Lettres is meant, "the know- ledge of the arts of poetry, and oratory; sometimes, that the true Belles Lettres are natural philosophy, geometry, and other essential parts of learning; and sometimes, that they comprehend the art of war, by land and sea: in short, they are made to include all that we know, and whatever we please; so that in treating on the Belles Lettres, they talk of the use of the sacraments, &c.*" In a word, it were an endless task to attempt to enumerate all the parts of literature that different learned men have comprehended under this title. The same indecision is to be found in the term *humanity* or classical learning; under which they include at pleasure, either more or less of the preparatory parts of learning, as grammar, rhetoric, &c. which are taught at schools, or in colleges, to such as are intended for the study of the superior sciences. In the midst of this uncertainty, it seems to be lawful for a private foreigner, who dwells at two hundred leagues distance from Paris, and is much embarrassed by so many different respectable authorities, to fix for himself the true import of the term; provided, however, that he humbly acknowledge his error whenever any master of the French language shall prove, by well-established usage, that he is wrong.

II. W&

---

* Rollin's method of teaching and studying the Belles Lettres.

II. We comprehend, therefore, under the term belles lettres, all those instructive and pleasing sciences which occupy the memory and the judgment, and do not make part, either of the superior sciences, of the polite arts, or of mechanic professions, &c. To these we consecrate this third volume of the analysis of the sciences; and we trust that we shall not omit any of them that ought naturally to be here included: for we hope, that memory and judgment will serve us as companions and guides in this long and difficult career.

III. All that relates to history or philology, requires at first, nothing more than sight and memory. In our earliest years those faculties are in their greatest vigour; all objects that then present themselves make the most lively and lasting impressions: the memory seems to trace on a young mind all those sciences which it is capable of comprehending, with indelible characters. The discerning faculty is formed more slowly; the mind requires a longer time to attain the capacity of distinguishing those objects that are presented to it by the sight or the memory. The judgment, or understanding, requires still more time to combine those objects, to compare them with each other, to draw from particular inferences general conclusions, to form them into systems, and to reduce them into sciences. Lastly, the genius, or inventive faculty, by aid of the senses, the memory

mory and the judgment, creates, produces, or discovers, either new truths, or undiscovered combinations, or brilliant comparisons, and striking images. This appears to us to be the natural progress of the faculties of the human mind, and by this progress man is conducted in the career of his studies. He should begin, in his early days, to apply to those sciences that exercise the memory; proceed to the forming of the discerning faculty; then elevate his mind to those superior sciences that occupy the judgment; and at length launch forth into the sublime regions of the polite arts; which are the produce of a well stored memory, an enlightened judgment, and a fruitful genius.

IV. The peculiar employment of childhood should be the learning of languages: for they are the instruments with which his mind is to work. To the beginning of youth, should be given a rough draft of the principal sciences of the memory, such as contains only facts, dates, and axioms: a sketch, for example, of history, a kind of gazette of simple events, without inferences or reflections, moral or political, without characters, and without ornaments. In the dawn of manhood, while the young student is preparing for the university, he should make himself a thorough master of logic, or the art of reasoning: he should then likewise acquire some tincture of the philosophic sciences; and make a second, more comprehensive, and more

rational

rational course in history. Now opportunities should also be given him of making some essays of his genius, that it may be conjectured of what future productions he may be capable. The university will furnish him with the necessary instructions in the superior sciences, and he will at last advance to the practice of the polite arts: he will invent, improve, produce; he will become at once a learned man, and a resplendent genius; even a Leibnitz, if providence shall permit.

V. History ought in a peculiar manner to be the study of every one, who would attain a liberal education; as it is a general storehouse for all the sciences, and a school for all the virtues. Whoever is appointed to instruct the children of princes, of the nobles, or principal inhabitants of the land, should endeavour, in the first place, strongly to impress on their minds a chronological series of all the remarkable events that are recorded in history, from the creation of the world down to the present day; making them well observe at the same time the several synchronisms, or the various events that have happened at the same period in different parts of the world. By these means he will open in their minds a repository, where every particular event may hereafter be ranged in its proper place; for, otherwise, without this, history would present a mere chaos to the memory, without order or connexion. When the student has thus

thus acquired a ready knowledge of chronology, he may undertake, with his tutor, a complete and rational courfe of hiftory: and there Clio fhould pluck for him the golden apples of the garden of the Hefperides. The animated and ftriking pictures of hiftory offer two forts of examples, the one to imitate, and the other to avoid. It is the bufinefs of an able inftructor carefully to point out, in the annals of all nations, thofe facts and characters that muft infpire their pupils with admiration or horror; and confequently excite in their minds a defire to imitate their virtues, and avoid their vices. The portraits of the truly great, as well as the tyrants of antiquity, when lively drawn, muft ftrongly affect the young ftudent; for they will feem to fay: " Future generations, princes,
" heroes, ftatefmen, fcholars, philofophers!
" Providence, for our greater reward, or more
" exemplary punifhment, has placed our ftatues
" in this gallery, to ferve as amiable or detefta-
" ble models to future ages. Emulate our vir-
" tues, and have a juft abhorrence of our crimes.
" Know that your real characters, that your
" actions, however abfurd or unjuft, and with
" whatever veil you may cover them, or under
" whatever mafk you may difguife them, will,
" like ours, ftand naked before pofterity. The
" piercing public eye will penetrate the moft
" fecret folds of your hearts. A thoufand
" fagacious obfervers continually furround you,
" and a thoufand pencils are conftantly ready to
" paint

" paint you to posterity, such as you really are.
" History flatters not: it is the witness, not the
" adulator of mankind."

VI. We must here make a few observations on the degree of credibility that a rational mind should give to the truth of history, or, in other words, on *historic faith*. No act or event can possibly happen, but such as is the result or produce of human actions, or the effects of nature: all actions must therefore arise from situations, circumstances or relations. We may be well assured, that all human actions, however extraordinary and wonderful, never have been, nor ever can be supernatural or miraculous; except those signal miracles only which God vouchsafed to operate, in order to establish the Judaic and Christian religions; and of which they are the foundations. These objects of our religious faith, of our piety and profound veneration, are as much above our weak comprehension, as sacred revelation is above philosophy, or mere human reason. It is with a lively, evangelic faith, that we are to acknowledge the truth and evidence of these facts. The historic faith on the contrary is, if we may use the expression, strictly argumentative. It examines, it doubts; and here doubt is the beginning of wisdom, for, as abbé Vallemont has very justly observed, there is no merit, either before God or man, in a stupid credulity.

VII. We

VII. We should take due care, therefore, not to push our historic faith so far as to believe all the prodigies, all the fables and extravagancies that are related by profane history, and especially that of the ancients. It would certainly be ridiculous to doubt that there have been such princes as Cyrus, Alexander, and Cæsar, and that they were great conquerors: but it would be still far more absurd to give credit to all the marvellous stories that have been related by historians: it would be madness to believe that Romulus and Remus were suckled by a wolf; that Numa Pompilius held an intercourse with the nymph Egeria; that the head of Ancus Martius burned in the Capitol; that Curtius threw himself into a gulph; or that the gods spoke by the means of oracles. Is it not ridiculous enough to see, in the eighteenth century of Christianity, a learned, elaborate and very serious dissertation, to prove that the oracles did not cease to speak at the coming of Jesus Christ; when it is evident to every man of any knowledge, that there never was any such beings as Jupiter or Apollo, and consequently that they never did speak? Such subjects as these ought to be ranked with the stories of giants, or the Tale of a Tub; and, whenever we meet in profane history with like accounts of prodigies and miracles, historic faith, or rather human credulity, should cease, and the sensible part of mankind should reason thus: either the gods were to blame so to dispose the order of nature,

nature, that it is not capable of producing the complete felicity of created beings, and especially of mankind, or else those gods were guilty of an absurdity, by interrupting the established order of nature, to produce effects, that might have been produced by merely following that eternal order. It is to be observed here, that we are now speaking of the gods of paganism only.

VIII. Historic faith is moreover founded entirely on human testimony, and that foundation is unfortunately very weak. What assurances have we, that the witnesses of events have never been deceived? or even that they have never been willing to be deceived? The same, and still more may be said of historians, who have been very rarely witnesses of the facts they relate, but have taken them merely from report. Now, if we suppose these facts to be certain, we must conclude, that these witnesses and historians were angels; for it is not in the nature of man to be infallible. The more witnesses likewise any prodigy has, for the most part, the more reason there is to suspect it: for the multitude are constantly inclined to deceive themselves; are fond of the marvellous, and drown the voice of the small number of the discerning part of mankind. We have seen the miracles of the blessed abbé Paris, that were attested by thousands of witnesses, whose veracity was indisputable, and yet they have at last been proved to be nothing more than artful impostures.

IX. The

IX. The imperfection in the frame of man, the weakness of his discernment, and the errors of his judgment, on one side, and the strength of his passions on the other, render his testimonies constantly equivocal and suspicious. Hear the accounts of two general officers that have been in the same battle; read the gazettes that relate the events which have happened in our own days, and frequently before our eyes, and judge how far you can depend upon the real truth of those facts. This being the case, you may easily determine what degree of credit is to be given to those marvellous relations, which are supposed to have happened among nations less enlightened than we are; in those ages, when learning was quite in its infancy, before printing was invented, and when the propagators of false reports stood in no dread of the severity of criticism. Let these and many other reflections, that we shall pass over in silence, set due bounds to your historic faith.

X. The passions likewise, to which human nature is liable, constantly cast a veil over the truth. It is an ancient saying, that an historian ought to have no religion, and no country. He is however, constantly, either a friend, or enemy of the prince or hero whose history he relates; he is prejudiced for or against a country, a people, a religion, a party or government. Passion continually guides his pen. We cannot read, without indignation, all that Tacitus writes a-
against

gainst Tiberius, whose professed enemy he was. Let Tiberius perform the most innocent, most just and honourable actions, Tacitus would find means to make them appear odious; though he frequently did it in a very awkward manner. Thucydides, Xenophon, and Josephus, were excellent historians; but if those people, who were the enemies of the Jews and Greeks, had found historians of equal ability with their antagonists, it is likely that the actions of the several heroes would have been set before us in very different lights. Notwithstanding the respect that is due to the fathers of the church, we cannot say that they were entirely free from passions. They gave to Constantine the surname of *Great*, who was doubtless one of the greatest dolts that ever existed; but he was a friend and protector of the Christian priests. The emperor Julian they represented as a monster, and a man of mean abilities; whereas he was one of the greatest men that history has recorded, his unfortunate apostacy excepted. Judge, after this, of the credit that is due to historians.

XI. The statesman and the scholar, the man of the world, and the man of genius, nevertheless, will and ought to make himself acquainted with history. He ought even to know it in the manner it has been transmitted to us, with all its fables, errors and falshoods. He ought to know, for example, all that the ancient historians have related of the labours of Hercules; of the expedition

expedition of the Argonauts; of the siege of Troy, &c. &c. though he do not give the same credit to these as to the gospel. It is of little import to us, whether these relations be true or not, either in substance or in circumstance; it is sufficient that we know in what manner history relates them. These marvellous stories even sometimes furnish assistance, pleasing ideas and allusions, to poetry and eloquence. The strict veracity of facts does not appear to become interesting to us, but in proportion as history approaches those ages that immediately precede the present; for the titles, the possessions, and pretensions of modern princes and nations, are entirely founded on these historical facts, and on the minutest circumstances that have attended them. The real influence of these facts and events on the interests of modern nations, can go very little further back than the time of Charlemagne. The principal points are, to determine in what state that monarch found Europe; what were then the rights of the people; after what manner he conquered them; by what method he established the western empire; what rights he thereby acquired; and what are the revolutions that have happened in the world from that period down to the present day.

XII. It is therefore from this famous epoch, that it concerns us thoroughly to know the veracity of facts, and of all their circumstances.

Those

Those of the preceding ages being more the objects of curiosity than utility, we shall leave them to the learned researches of critics, antiquaries, and commentators; acknowledging the obligation we have to their laborious inquiries. We shall say nothing here of the study of the other parts of historic and philologic science. That only requires, as we have already observed, good eyes, a just discernment, and a happy memory. What remarks may be necessary relative to that matter, we shall make in the course of our analysis of those sciences.

XIII. The love of truth obliges us to make here one observation, and which we do at the risk of offending, and regardless of the consequence. Every man who would acquire a true knowledge of the historic sciences (and frequently of the philosophic also) should learn them from such works as are wrote by Protestants. The inquisition of the church of Rome strikes all catholic writers, and especially historians, with a wretched timidity, that constrains them to disguise the truth, or at least to suppress it, and be silent on all those matters that can in the least affect their religion. In all such facts as relate to the origin and increase of the hierarchy, those authors are to be altogether suspected; especially when they belong themselves to the ecclesiastical state, and their fortune visibly depends on the court of Rome. We will defy any one to produce a single work of this kind,

in

in which we cannot point out visible marks of this unhappy truth; and which we find ourselves obliged, however unwilling, to declare in this place. The inconvenience is greater than is easily imagined.

XIV. And now, ye Studious Youth, who seek to inform yourselves by this abridgment of the course you should pursue in the study of the sciences, constantly remember, that theory alone, however perfect it may be, will perpetually remain a barren knowledge; that history, especially, should direct you to a sagacious conduct, should inspire you with a love of virtue, and with an aversion to folly and to vice. Be not therefore content with knowing much, but let your knowledge be the guide to your talents; for, in a word,

   Omnia tendunt ad praxin.

CHAP. II.

MYTHOLOGY.

I. THE word *mythology* is a Greek compound, that signifies *a discourse on fables*; and comprehends, in a collective sense, all the fabulous and poetic history of pagan antiquity. It follows therefore, that this science teaches the history of the gods, demi-gods, and fabulous heroes of antiquity; the theology of the pagans, the principles of their religion, their mysteries, metamorphoses, oracles, &c. By this definition, it appears sufficiently what are the objects of which we are to treat in this chapter.

II. If we well consider the matter, we shall find, that there were, in pagan antiquity, three different religions. First, That of the philosophers, who treated metaphysically of the nature, the attributes, and of the works of the Supreme Being. They endeavoured to discover the true God, and the manner in which he ought to be worshipped.

It

It is not wonderful, that these men of exalted genius should in some degree ridicule, in their works, the two other positive religions, and those gods on whom they were founded; at the same time that they outwardly professed the established religion, in order to preserve the peace of society, and to avoid the persecutions of the legislature, and the insults of the populace. For in fact, was it possible for them to believe the pagan fables? Must they not foresee, that their religion would one day give place to another, while their own works would pass with their names to the latest posterity? And could they suffer the thought, that their reputation would be tarnished in the eyes of that posterity, by having it imagined they believed such idle tales as were broached by the priests of their times? Could Plato, Socrates, Seneca, and Cicero, be unconcerned for their fame among future generations, and future philosophers? And what should we at this day have said of those great men, had they been so political, or hypocritical, as to have entirely concealed their sentiments with regard to these matters?

III. The second religion was that of paganism, which was the established religion of all the ancient nations, except the Jews. This was the doctrine that was taught by the priests, and protected by the sovereigns. Its dogmas were demonstratively false, but not always so absurd as may at first appear, especially if we annex (as I think

MYTHOLOGY. 17

think we fhould) to the divinities, and to the religious ceremonies of the pagans, a fenfe that is frequently myftic, and always allegoric; if we remember, that the firft heathens deified thofe great men to whom the reft of mankind were indebted for any fignal benefits, as Jupiter, Apollo, Ceres, Bacchus, Hercules, Æfculapius, &c. in order to induce others, as well of the prefent as future ages, to reverence and to imitate them. Would not an ancient pagan, if he were to return upon the earth, have fpecious arguments, at leaft, to fupport his religion, when he faw weak mortals beatify or canonize, merely by their own authority, other weak mortals (frequently mere pedants) and place them in heaven, without the permiffion or approbation of the Supreme Being? Happy is it for mankind, when at different times fagacious pontiffs purge the calendar, and the brains of the people, from a herd of pretended faints, and prevent them, at leaft after their death, from doing injury to fociety, by interrupting the induftry of the laborious inhabitants with keeping their feftivals.

IV. The third religion was idolatry, or the religion of the populace. For the common people, born to be deceived in every thing, confounding in their imaginations the ftatues of the gods, the idols of their divinities, the emblems of their virtues and of religious worfhip, with the gods, divinities, virtues and worfhip themfelves, adored thefe images, and proceeded to

VOL. III. B extravagancies

extravagancies the moſt ridiculous, and frequently moſt criminal, in their ceremonies, feaſts, libations, ſacrifices, &c. It is to be feared, that, as long as there are upon the earth men of our limited capacities, this triple religion will conſtantly ſubſiſt under different forms; and we are much deceived, if it may not be found under the empire of Chriſtianity itſelf, notwithſtanding the purity of its doctrine. It will be eaſily conceived, that it is not of the religion of philoſophers, nor that of the populace, of which we are to treat in this chapter on Mythology; but of that which ſubſiſted under the authority of the magiſtracy and the prieſthood; and conſequently of paganiſm in general.

V. As far as we are able to judge by all the ancient authors we have read, the pagans adored the Sovereign Lord of the univerſe under the name of *Fate* or *Deſtiny*, (Fatum) which we muſt not confound with *Fortune*, who was regarded as a ſubaltern divinity. Jupiter himſelf, all the gods, every animated being, the heavens, the earth, the whole frame of nature was ſubſervient to Deſtiny, and nothing could reverſe its decrees. This divinity was ſo highly adorable, as to be above all rank, and was regarded as too ſupreme to be repreſented under any ſenſible image or ſtatue, or to have any temple erected for its worſhip. We do not remember to have read, that any ſacrifice was ever offered to this Deſtiny, or that any temple or city was ever dedicated to its name.

name. We are almoſt inclined to think, that the pagans were ſenſible, that the temple and the worſhip of the God of gods ought to be in the heart of man. Mention is made, indeed, of a temple that was dedicated to the unknown God, but we are ignorant whether or not Deſtiny were thereby meant. We muſt not confound this Deſtiny, moreover, with the goddeſs of chance, of which there are ſome antique ſtatues that repreſent her in a recumbent poſture, and playing with little bones; for this was nothing more than an invention of ſome ſtatuary.

VI. After this general and philoſophical idea of the Supreme Being, comes the poſitive religion of the pagans. This was entirely founded on fable, which took its riſe either from ancient traditions, or hiſtorical events, altered or augmented by the imaginations of the poets, by ſuperſtition, or by the credulity of the people; or elſe it conſiſted of allegoric or moral fictions. A crowd of writers, and among the reſt Noel le Comte, (Natalis Comes) the abbots Bannier and Pluche, &c. have made many reſearches into the origin of fable: and they think they have diſcovered its ſource, 1. in the vanity of mankind; 2. in the want of letters and characters; 3. in the deluſive eloquence of orators; 4. in the relations of travellers; 5. in the fictions of poets, painters, ſtatuaries, and dramatic writers; 6. in the diverſity and uniformity of names; 7. in the ignorance of true philoſophy;

8. in

8. in the foundation of colonies, and the invention of arts; 9. in the defire of having gods for our anceftors; 10. in the imperfect or falfe interpretation of the holy fcriptures; 11. in the ignorance of ancient hiftory; 12. in a like ignorance of chronology; 13. in that of foreign languages; 14. in the tranflation of the religion of the Egyptians and Phœnicians into Greece; 15. in the ignorance of geography; and 16. in the belief that the firft people had of the intercourfe of gods with men. It is certain, that all thefe matters taken together are fufficient to produce many thoufands of fables; are more than fufficient to enable us to deceive ourfelves and others, and to give rife to infinite reveries. But we fhould take care how we draw from thefe fources demonftrations that might be ufed, by infidels, as arguments to overthrow the hiftory of the Jews; a people the moft ftupid, moft credulous, and oftentatious of all others. In the mean time, the pagan philofophers themfelves afferted, that it was a god who invented the fable: fo much they were convinced of its ingenuity, and of its ftrong tendency to inftruct mankind in their duty.

VII. Mythology therefore, when properly treated, begins with making learned refearches into the real origin of fable, of paganifm, and of that idolatry which was its confequence. It recurs for this purpofe even to the beginning of the world; and after finding that Laban, the
father-

father-in-law of the patriarch Jacob, was a maker of idols, and that he had his little images, or houshold gods, which he formed of baked earth, and which shows, that idolatry existed in the greatest antiquity, it then explains *cosmogony*, and *theogony*, or the belief that the first inhabitants of the earth entertained of the creation of the universe, and what the pagan theology taught of the genealogy of their false gods. It begins with the tradition of the Chaldeans, a people so ancient, that Nimrod was their first king; but at the same time, so credulous and superstitious, that we may regard them as the authors of all those fables, and the propagators of all those visions, that have since blinded human reason. According to this tradition, a monster named *Oannes*, or *Oes*, half fish and half man, sprang from the sea, before the chaos was completely dispersed, and gave laws to the Chaldeans. A woman, called *Omorka*, reigned over all the earth. *Bel* cut her in two, and made of one moiety the heavens, and of the other the earth. They likewise invented the two primitive beings, of which the good one, who was named *Oramasdes*, had the direction of heaven, and the other called *Arimanius*, that of hell.

VIII. The science of mythology then teaches the theogony of the Phœnicians; concerning whom it draws great lights from Sanchoniathon, a priest of Beryte, who lived before the Trojan wars, more than four hundred years before Hesiod

siod and Homer, and of whom Eusebius has preserved considerable fragments. From thence it passes to the theogony of the Egyptians, of whom *Thot* or *Thaut*, the founder of that nation, was likewise, they say, their first historian, that Sanchoniathon even copied from him; and of whom we find many relations in the Greek historians, especially in Herodotus, Diodorus Siculus, and in Eusebius of Cæsarea. It then examines the theogony of the Atlantides, who dwelt on the western part of Africa, and of whom Diodorus alone has preserved any account. From thence it proceeds to the theogony of the Greeks, which is far better known to us, as we find accounts of it, more or less particular, in numberless Greek and Latin writers. This theogony had the same foundation as that of the Romans; the latter having only extended it by adding to the Greek divinities certain gods or demi-gods, formed of their heroes, and certain symbolic and allegoric divinities, which mythology explains at the same time; and it is on this occasion, that it enters into a particular explication of the cosmogony and theogony of Ovid; whose book of metamorphoses contains as copious descriptions as we could desire of the fable of the ancients: what was their belief concerning the habitations of the blessed after their death, or of the Elysian fields; as well as of their hell or Tartarus; of the dog Cerberus; of the ferryman Charon; of the Furies; of the four rivers, Cocytus, Lethe, Phlegethon and Styx, which water the
Tartarian

Tartarian regions, &c. The learned have likewise made many inquiries, and many ingenious discoveries concerning the theogony of the ancient Germans, Celts, the Scythian and Hyperborean nations. In the last place, this science furnishes great lights on the theogony of the Bramins, the Troglodytes, the Indians, the Chinese, and even the Americans; all which it concludes with a regular and minute examination of the pagan theology, and particularly that of the poets.

IX. All these matters being well digested in the minds of those who would make a regular study of pagan theology, they continue their researches into the time, the epoch and place of the real origin of paganism and idolatry, and they prove that the pagans began by adoring the heavenly bodies, the stars and planets. They next examine into the progress of idolatry, what were the temples of the pagans, their altars, their enclosures, their sacred groves, their asylums, the idols and statues of their deities; in what manner they were represented, what were their sacrifices, the victims that were offered, what were the sacred vessels, the censers and other instruments that were used in the sacrifices, libations, and other religious ceremonies; concerning the priests, priestesses, and other attendants on the service of each divinity: what were the festivals that were celebrated among the Greeks and Romans, as
well

well as among the Orientals: what the days of penitence and supplication, the feasts of the gods or lectisternia, their invocations or incantations, and exorcisms, the religious ceremonies observed at laying the foundations of cities, &c.

X. Divination, or the prediction of future events, a weakness that has at all times possessed the human mind, forms also an important article of pagan theology. It is therefore in this place, that mythology considers the nature of *Oracles*, and in particular, 1. The oracle of Dodona, the most ancient of Greece. 2. That of Jupiter Hammon or Ammon, in Lybia. 3. That of Jupiter Philius. 4. That of Apollo, both of Heliopolis. 5. That of Apollo of Delphos. 6. That of Trophonius in Bœotia. 7. That of Venus of Aphaca, a country between Byblos and Heliopolis, situate on a small lake; and a great number of other oracles of less note, dispersed over Greece and other countries. It also examines in what manner these oracles gave their answers, the ceremonies that were observed in consulting them, the frantic emotions of the priestess Pythia on her tripod; and those of other priests. It then endeavours to determine if there ever were in fact any Sibyls, which, whatever has been said, is still very doubtful; it draws, however, from all the sources of antiquity, a kind of history of these Sibyls and of their prophecies. It next

next passes to the examen of the nature of auguries, auspices, haruspices, presages, prodigies, and phenomena, of expiations and ablutions, of the magic and astrology of the ancients, &c. Whoever has thoroughly studied all these objects, is fully provided with the preliminary knowledge that is necessary to enable him to proceed steadily and securely through the darkness of ancient mythology, and he may thereby advance more confidently to the examination of the nature of the pagan divinities themselves.

XI. The celebrated treatise of Cicero *de natura deorum* will here furnish great lights; but modern authors who have treated on these matters, have not been contented with this alone; they have, so to say, extracted the essence of all antiquity, of which they have formed systems; but unluckily these scarce ever agree with each other. As philosophers, it is of very little importance for us to know what was the nature of these gods, seeing we know that they were merely fabulous: but as historians and antiquaries, it concerns us to know what was the nature that was attributed to them in general, and in particular, what were the origin, genealogy, rank, functions, authority and operations, that were attributed to each divinity; and it is on these matters that we have still some remarks to make.

XII. The

XII. The gods of the ancient Greeks and Romans were all either *Dii majorum gentium*, or *Dii minorum gentium:* that is, of the first or second order. The former were also called *confentes, magni confultores,* &c. According to Ennius they were twelve in number, and are included in these verses:

Juno, Vesta, Minerva, Ceres, Diana, Venus, Mars, Mercurius, Jovis, Neptunus, Vulcanus, Apollo.

To these were added eight others under the title of *felecti*, which were Sol, Luna, Tellus, Genius, Janus, Saturnus, Liber, and Pluto. The second order, or *minorum gentium*, were called Adscriptitii, Medioximi, Minuscularii, Putatitii, Indigetes, Semones, &c. the principal of which were Æsculapius, Bacchus, Castor, Fauna, Hercules, the Lares or Penates, Pollux, Quirinus, Semo Sancus or Dius Fidius, &c.

XIII. According to the second division, all their divinities were classed into, 1. Celestial gods, 2. Terrestrial gods, 3. Sea gods, and 4. the Infernal deities, or *Inferi*. The celestial gods were Jupiter, Juno, Apollo, Aurora, Cupid, Cybele, the Graces, Hebe, Iris, Luna, Mars, Mercury, Minerva, Nemesis, Saturn, Themis, Venus, &c. The terrestrial gods were Æolus, Astræus, Astræa, Ceres, Diana, the Fauni, Feronia, Flora, Janus, Momus, the Muses, Pales, Pan, Pomona, Priapus, the

the Satyrs, Silenus, Silvanus, the god Terminus, Vesta or Rhea, Berecynthia, Vulcan, Harpocrates, &c. The sea gods were Neptune, Amphitrite, Thetis, Canopus, Glaucus, Ino, the Nereids, Nereus, Oceanus, Palæmon, Triton, &c. The infernal gods were Pluto, Proserpine, Charon, Minos, Æacus, Rhadamanthus, the Furies, Death, Night, the Fates, Plutus, &c.

XIV. The third division ranged the divinities according as they presided, 1. Over the pregnancy of women (*Prægnantium*.) 2. At parturitions (*Parturientium*.) 3. At births (*Nascentium*.) 4. At adulteries. 5. At marriages: to which they added, 6. Dii morales, or moral gods, and 7. Funeral gods. The gods of pregnancy were Pilumnus, Intercidona, and Deverra: the gods of parturition, Juno, Lucina, Diana, Egeria, Prosa, Postverta, Menagenata, Latona, the gods that were called Nixi, or of labor, &c. The gods of birth were Janus, Opis, Nascion, Cunina, Carmenta, Vaginianus, Levana, Rumia, Potina, Educa, Ossilago, Carnea, Nundina, Statilinus, Fabulinus, Paventia, &c. The gods of adultery were Juventus, Agenoria, Strenua, Stimula, Horta, Quies, Murcia, Adeona, Abeona, Voluptas, Orbona, Pellonia, Numeria, Camoena, Sentia, Angerona, Heres, Martea, Laverna, the god Averruncus, Consus, Catius, Volumnus and Volumna, Honorius, Aius Locutius, &c. The nuptial gods were Diana, Domiduca,

Domiduca, Domitius, Hymenæus or Hymen, Jugatinus, Jupiter perfectus, Juno perfecta, Juno cinxia, Juna unxia, Lucina, Manturna, Mutinus, Dea Mater prema, Suada, Thalaffius, Venus, &c. The moral gods were called Virtus, Honor, Fides, Spes, Juftitia, Pietas, Mifericordia, Clementia, Pudicitia, Veritas, Mens, Concordia, Pax, Salus, Felicitas, Libertas, Pecunia, Rifus, Invidia, Contumelia, Impudentia, Calumnia, Fraus, Difcordia, Furor, Fama, Fortuna, with all their epithets good or bad, Febris, Pavor and Pallor, Paupertas, Neceffitas, Tempeftas, Silentium, &c. The funeral gods were Pluto, Libitina, Nænia, Death, the Fates, &c.

XV. Hefiod indeed pretends that all thefe gods derived their origin from chaos, but we have already pointed out more juft fources. It is almoft incredible to what a prodigious number the fuperftition and weaknefs of the Greeks and Romans multiplied thefe divinities; there have been thirty thoufand of them enumerated. It will not be expected that we fhould here attempt to defcribe them, nor will it be remarkable if we have forgot to mention even fome of the firft rank. Although vaft as this company of gods is, mythology does not omit to trace the hiftory of the greateft part of them, as it is taught by paganifm; and they who are defirous of particular information in thefe matters may confult with advantage the theogony of Hefiod,

the

the catalogue of Apollodorus, the metamorphoses of Ovid; the fables of Hygina; Lylii Gregorii Gyraldi Syntagma de Diis Gentilium, the mythology of Natalis Comes; the books of Gerard Vossius de Idolatria Gentilium; Johannis Boccatii Genealogia Deorum; the Pantheon of Pomey; the history of heaven by abbé Pluche; the historic explanation of fables, by abbé Bannier; and numberless other works of the same kind in all languages.

XVI. There were still many other distinctions, of which the pagans made use to mark their rank, the functions and nature of their several divinities. For example, the goddess Vesta, or the mother of all the gods, was adored by all people in general. Mars, Bellona, Victoria, Fortunata, &c. assisted all parties. The *topical gods*, on the contrary, were adored in particular countries only; as Astarta in Syria, Derceto and Semiramis among the Assyrians. Isis and Osiris by the Egyptians; Quirinus at Rome, &c. The title Semones, which was given to a certain class of divinities, was doubtless derived from Semi-homines, that is, demi-men, and signified the same as semi-dii, or demi-gods. These were monarchs and illustrious heroes, or those great men who were the founders of cities and nations, that were deified by way of apotheosis. Pythagoras had taught the Chaldeans the doctrine of transmigration, and that after their death, those who were virtuous, would be elevated

elevated to the rank of divinities. This doctrine was adopted by all the pagan world. The apotheosis, after they had erected temples and altars to the new gods, was celebrated with much solemnity. In the last ceremony, an eagle was fixed on the catafalk, or funeral pile, on which was placed the image of the hero, and when the pile began to burn, the eagle was let loose, who, mounting into the air with the flames, seemed to carry the soul of the departed hero up to heaven.

XVII. Mythology informs us also, who those persons were that antiquity regarded as the children of the gods, such as Theseus, Hippolytus, Paris, &c. what the pagans believed, with regard to the nature of their Genii and Demons, of their Dryades, Hamadryades, Nymphs, Tritons, Sirens, Fawns, Silvans, Centaurs, and other subaltern divinities; and in this manner it explains all the systems of the positive religion of the Greeks and Romans. They who are desirous of extending their knowledge of paganism still further, of knowing the dogmas of each particular people, what were their gods, and the various manners in which they were worshipped, such as Apis, Isis, Osiris, &c. the adoration of crocodiles and onions, &c. among the Egyptians, must study the different theogonies of these people, and notwithstanding all the informations which ancient and modern authors afford, this study is yet boundless, and attended with

with many difficulties and uncertainties. Though it appears demonstrative, that the origin of paganism, and of idolatry in general, was derived from the Chaldeans, from whom the Egyptians drew that doctrine which they after transmitted to all other nations; and consequently that the primordial divinities were the same, under different denominations, among all the idolatrous nations of the earth.

XVIII. The nature of this work will not permit us to descend to further particulars. But to give our readers an idea of the manner in which mythology treats its subjects, and of the method that should be observed in studying fable, or the history of the gods of antiquity, we shall here give, by way of example, a cursory description of Parnassus and its inhabitants.

Parnassus was a mountain of Phocis, that had two summits, one of which was called Tithoreus, and the other Hyampeus. Others say, that one of these hills was named Helicon, and the other Cytheron, and that it is an error to imagine, that Helicon was a mountain of Bœotia. However that be, this double hill was consecrated to Apollo and the muses, who there held their usual residence. According to fable, there had been a remarkable combat on this hill, between Helicon and Cytheron. Whoever slept on Parnassus, when he waked, became a poet. Apollo had there

there a temple. There also was the fountain Castalia, into which Apollo had metamorphosed a nymph that he loved, and had given to its waters the power of making all who drank of them poets. At the foot of Parnassus flowed the river Hippocrene, that had the same virtue; and the source of which was opened by a stroke of the foot of the horse Pegasus. This river nourished a great number of swans, that were regarded as sacred. Pegasus was a winged horse, that belonged to Apollo, and grazed on the summit of Parnassus. He sprang from the blood of Medusa, when Perseus cut off her head, which was placed among the stars. Such was the delicious abode of Apollo, the son of Jupiter and Latona, who was born, with his twin sister Diana, in the island Delos. He killed the Cyclops who forged the thunder bolts with which Jupiter had overthrown his son Æsculapius; but for that presumption, he was forced to leave heaven, and to become an inhabitant of the earth. He guarded the oxen of Admetus; he aided Neptune to build the walls of Troy; and Alcotheus in forming the labyrinth. He killed the dragon or serpent Python. He invented music and physic; and was honoured as the god of poets and physicians. He was represented as a young man, without a beard, his head surrounded with rays, and bearing in his hand a bow, or a lyre. As the ancients denoted the sun by the name of Apollo, they sometimes represented him also as seated in a chariot, drawn by two white horses,

preceded

preceded by Aurora and the ſtar Venus: Phaeton his ſon, being deſirous of conducting theſe horſes, was thrown into the ſea. Apollo was alſo called Phœbus, Titan, and Sol. He is known to have had amours with Arſinoe, Corycia, Meloene, Cyrene, Mantho, Sinope, Calliope, and others; by whom he had Delphe, Naxe, Miletus, Arabe, Garamas, Sirus, Linus, Orpheus, and other children. He had peculiar honours paid him in the Pythian games at Delphos, and in the ſecular games at Rome.

XIX. The Muſes were the companions of Apollo in his rural abode. They were likewiſe called the learned ſiſters; as alſo the Camœnian, Heliconian, Parnaſſian, Aonian, Pierian, Pegaſian, Aganippian, Theſpian, Libethrian and Caſtalian ſiſters. They were the daughters of Jupiter and Mnemoſyne, and were regarded as the goddeſſes of ſciences and arts in general. There were nine of theſe muſes, to whom they attributed, 1. to Clio, hiſtory, 2. to Melpomene, tragedy, 3. to Thalia, comedy, 4. to Euterpe, flutes and other pneumatic inſtruments of muſic, 5. to Terpſichore, the harp and the dance, 6. to Erato, the lyre and the lute, 7. to Calliope, heroic verſe, 8. to Urania, aſtronomy, and 9. to Polyhymnia, rhetoric and eloquence. The Graces alſo ſometimes quitted Venus to pay their court to Apollo.

XX. Such was the idea they entertained of Parnaſſus and its inhabitants. There is no doubt but that under theſe fabulous repreſentations, theſe ſenſible images, were concealed allegoric and moral meanings; nor can it be denied but that their method of cultivating the arts and ſciences, by this manner of expreſſing their ideas, was as ingenious and pleaſing as it is poſſible to imagine. Every other ſubject that paganiſm embraced, it treated with the ſame genius and in a manner equally pleaſing; and though that religion was altogether fallacious, yet we muſt allow that it was extremely well calculated to promote the polite arts, by thoſe refined, noble, graceful, brilliant images, by thoſe charming ſubjects which it conſtantly preſented, and which it ſtill offers to the poet, painter, ſculptor and every other artiſt.

XXI. But this was not a power ſufficiently ſtrong to ſecure paganiſm againſt that viciſſitude, that decline and diſſolution, which finally attends all the productions of this world. This religion, which had ſubſiſted near five thouſand years, and almoſt from the origin of the human race, gradually declined in proportion as the lights of Chriſtianity and philoſophy illumined the minds of mankind. For though the pagan religion, and the fables on which it was founded, were pleaſing and favourable to the polite arts, they were not however calculated to ſatisfy the minds of philoſophers, nor to promote the real good of

mankind,

mankind, by securing their temporal and eternal happiness. It is even surprising, that so great a genius as the emperor Julian should attempt to revive the embers of paganism, which insensibly declined, and had received a mortal blow at the beginning of the fourth century by the emperor Constantine the Great. Julian employed all the resources of his imagination, of his eloquence, of his power, and even of his own fatal example, to revive it, but in vain. The fatal period of paganism was arrived, and nothing could save it from destruction: The furious Theodosius, to whom bigotted priests and historians have assigned the name of Great, totally overthrew it toward the close of the same century; destroyed those temples and altars which yet subsisted, dispersed its colleges and exterminated its priests. From that dire epoch, nothing of paganism has remained, except some ruins dispersed in the remote parts of the earth, and among people wretched and almost unknown; where this religion, once so flourshing and universal, is now degenerated into gross and disgustful idolatry.

CHAP. III.

CHRONOLOGY.

I. CHronology is the science that teaches *the method of measuring time and distinguishing its parts.* It is more difficult, than may at first appear, to determine the precise idea, and clearly to explain the nature of time. That ingenious and subtile impostor Mahomet has given in his Alcoran some traces of very refined ideas of this subject. But, leaving these metaphysical researches, we shall content ourselves with saying, that by time we here mean the duration and succession of created beings. To determine a fixed and sensible measure of duration, it is necessary to find some motion that is constantly uniform, which may serve as a scale for that measure. From the creation of the world, it has been observed that the courses of the heavenly bodies afford the most universal measure of motion to all the inhabitants of the earth. As it was originally imagined that the sun turned round the earth, his annual and diurnal revolutions were fixed on for the common measure of time; and by this measure they divided the duration

ration of beings into years, months, weeks, days, hours, minutes and seconds. It may seem strange to an astronomer, or chronologist, to read, in the first chapter of Genesis, that God did not create the sun, moon and stars till the fourth day, and that there were days and nights before there was any sun. But who can say what is there precisely meant by the word day? Moses, who lived about three thousand years after the creation, wrote the origin and history of the Jews. In order to which he recurred to the origin of all things: he began with the creation itself: but he wrote to men, and to men who were even less enlightened than we are, especially in matters of astronomy. He was therefore obliged to make use of expressions that were to them intelligible. The scriptures were moreover given to mankind to serve them as guides in matters of religion, and not to teach them astronomy; of which were they ignorant, they would be obliged to believe, for example, that the sun moves round the earth, and that it was stopped, though a thousand times greater than the whole terrestrial globe, by the desire of Joshua at Gibeon; and that the moon halted in the valley of Ajalon, &c. all which is directly contrary to the eternal laws of nature, and therefore, taken in the strict letter, cannot be true. But who knows what means Providence may have employed to produce these appearances? Without making further inquiry into these matters, let us acknowledge the goodness of the Holy

Spirit

Spirit that has vouchsafed to speak to mankind in a language adapted to their capacities, in pointing out the path that leads to eternal felicity; where those dark clouds which now surround the human understanding shall be dispersed, and it will then perhaps discover many of those positions to be errors which philosophers and astronomers now regard as axioms, or incontestable truths.

II. Since Copernicus has discovered that the earth moves in its orbit round the sun, it necessarily follows, that the measure of time arises from the motion of this our globe. But as chronology is founded on apparent astronomy, or on that part of it which considers the celestial bodies and their motions as they appear to our senses, and forms its calculations in consequence, all that we shall here say of its operations, will therefore relate to that part of astronomy which is regulated by appearances.

III. The term chronology, when taken in its full extent, has two objects that may seem to be in a manner two different sciences, but which have a natural connexion. The first is the measuring of time and its different divisions; now this part of chronology is regulated by astronomical calculation, and consequently makes a part of mathematics. And it is by this method that we are enabled to make complete calendars or almanacs. The second part of chronolgy
consists

consists in fixing the dates of all those events that are related in history, and of ranging them in the several divisions of time in which they occurred: and by this means chronology becomes one of the essential parts of history. This second part of chronology draws its principles from the first; but it has need of other supports, as of criticism, of the testimony of authors, of ancient coins, medals, inscriptions, &c. of such epochs in history as are incontestable; of eclipses of the sun and moon, and other astronomical observations, &c. We shall now make the analysis of chronology according to this natural division, and shall consider it from these different points of view.

IV. The time that the sun employs in going completely round the earth is called a day. We also call that time the sun remains above the horizon, day; and the time he is under it, night. As the sun's motion is flower when he is in the apogee than when in the perigee, it follows that the first sort of days, which are also called natural days, must be shorter in summer than in winter. The natural day is divided into four-and-twenty hours, the hour into sixty minutes, and the minute into sixty seconds. As the point of mid-day or noon can be observed, by means of the meridian, with the greatest precision, astronomers begin the day at that point, and count twenty-four hours in succession; which, when thus counted, are called astronomic hours. The common people, on the contrary,
begin

begin the day at midnight, and count twelve hours to mid-day, and from thence twelve hours more to midnight; and these are called European hours.

V. The ancient Arabs, and some other nations, began their day with the astronomers; but the Egyptians and Romans at the same time we do. The Italians and Chinese (as did also the Athenians) begin their day at sunset; and the modern Greeks, by the example of the Babylonians, begin it at sunrise. The hours therefore that are counted after the former method are called Italian, and the latter Babylonian hours: and in both methods they count twenty-four hours in succession. The Jews begin the day also at sunset: anciently they divided each day, whether long or short, into twelve hours, and the night the same. These unequal hours are called Judaic or planetary hours: the Judaic hours therefore are long or short, according to the duration of the day. The Chaldean scruple is the $\frac{1}{1080}$ part of an hour. The Jews, Arabs, and other oriental nations, make use of this division, and call these scruples *Helakim*. Eighteen Chaldean scruples are equal to one minute, and consequently 15 minutes are equal to 270 scruples.

VI. A week is the space of seven days. This division of time took its origin from the creation. It was adopted by the patriarchs and other Jews, and has passed from them to most other nations.

The

The Persians, however, do not count by weeks, nor do some of the Indian nations. We owe the names of the days to the Egyptians and astrologers, who have given to each day the name of that planet, which, according to them, reigns over the first hour of that day, beginning with Saturday. They therefore range the days as follows:

♄ *Dies Saturni*, - or - Saturday.
☉ *Dies Solis*, - - - - Sunday.
☽ *Dies Lunæ*, - - - - Monday.
♂ *Dies Martis*, - - - Tuesday.
☿ *Dies Mercurii*, - - - Wednesday.
♃ *Dies Jovis*, - - - Thursday.
♀ *Dies Veneris*, - - - Friday.

Christian astronomers and chronologists have preserved these signs of the Latin names in their almanacs; but we begin the week with Sunday *(Dies Solis)* the day that Christians consecrate to devotion, and to the memory of the resurrection of our Saviour; their week therefore ends with Saturday, or the day of the Jewish Sabbath. Sometimes they also mark the seven days of the week in the calendar by the first seven letters of the alphabet; thus,

A. Sunday.
B. Monday.
C. Tuesday.
D. Wednesday.
E. Thursday.
F. Friday.
G. Saturday.

Which

Which is of use in calculating the days, as each letter or sign, that is once adopted to signify any particular day, constantly denotes the same day throughout the year.

VII. A *solar month* is the space of time that the sun employs in passing through a sign of the Zodiac. The solar months are equal among themselves, and, according to the mean motion, each solar month is equal to 30 days, 10 hours, 29 minutes, and 5 seconds. But this kind of month cannot be used in the common affairs of life, as we can there only count by whole days. A *lunar month* is the space of time from one new moon to another. The duration of a lunar month being 29 days, 12 hours, 44 minutes, and 3 seconds, cannot, for the same reason, be observed in common life.

VIII. A *solar year* is the time in which the sun runs through all the twelve signs of the Zodiac, and is consequently composed of twelve solar months. But there are here two necessary observations to be made. The first is, that the solar year, consisting of 365 days, 5 hours, and 49 minutes, it cannot likewise be observed in common life; and great confusion would arise if the year did not constantly begin on the same day. The solar year, therefore, is reduced to 365 days only, and when the odd hours and minutes amount to a day, it is added to that year, which then consists of 366 days. The second observation is, that when 365 is divided by 12,
the

the quotient is $30\frac{1}{11}$; therefore, as the solar year consists of twelve months, seven of these months should have 30 days, and five 31; and when the year consists of 366 days, there should be six months of 30 days, and six of 31. But in our chronology a different method is observed. In the common year, of 365 days, the months of January, March, May, July, August, October and December, have 31 days each; those of April, June, September, and November, 30; and the month of February 28 days: but when the year consists of 366 days, February has 29 days: such a year is called Bissextile, or Leap-year, and the day that is added is called the Intercalary day. It is also necessary to observe, that as the time above 365 days consists of 5 hours 49 minutes, there will be in a century, beside the 24 intercalary days, a surplus of 5 hours and 40 minutes, which, in 400 years, will amount to 22 hours 40 minutes, or almost a day, which must therefore be also intercalated at the end of the fourth century.

IX. The *lunar year* is composed of 12 lunar months, and consists of 354 days, 8 hours, 38 minutes, and 36 seconds: consequently the difference between the solar and the lunar year, amounts to 10 days, 21 hours, 24 seconds. Chronology therefore demonstrates, by the aid of astronomic calculation, that, in a hundred lunar years, there must be intercalated about 53 months; unless we would have the beginning of the

the year run through all the seasons, and fall sometimes in summer, and sometimes in winter.

X. The *common Julian year* has 365 days, and the bissextile 366. The fourth year is always bissextile. The emperor Julius Cæsar, the reformer of the Roman calendar, fixed the solar year, by the advice of his astronomer Sossygenes, at 365 days, 6 hours, and consequently at 11 minutes more than the truth; and which produced, in a hundred years, a difference of 18 hours and 20 minutes. The Julian year was used throughout all Christianity till the year 1582, when pope Gregory again altered the calendar.

XI. The *common Gregorian year* consists, like the Julian, of 365 days, and the bissextile of 366. But as in a hundred years there can be only 24 bissextiles, at the end of four hundred years there will consequently be a surplus of 22 hours; Gregory therefore appointed the bissextile every fourth year, but at the end of the century he directed there should be three common years together, and has fixed the bissextile only at the end of the fourth century: which makes a difference with the true solar year of 1 hour and 20 minutes in 400 years, and consequently a whole day in 7200 years. On the other hand, the Gregorian year begins, in 400 years, always three days sooner than the Julian year. This difference had increased, from the time of the council of Nice to the pontificate of Gregory, to 10, and at the beginning of the present century, to 11 days. These 11 days have therefore been rescinded from the calendar, and this last reformation is called the New Stile, and has been adopted by all the nations of Europe.

XII. The names of the months, and the number of days they contain, are to be found in all almanacs. The Romans reckoned at first only 10 months, from whence came the names September, October, November, December. They had also a peculiar method of counting the days. The first day in each month they called the Calends. The calends were followed in the months of March, May, July, and October, by six Nones, and in the other months by four Nones. These Nones were also followed by eight Ides, and the rest of the days were called the Calends of the succeeding months; as appears by these verses:

Prima dies mensis cujusque est dicta Calendæ.
Sex Maius, Nonas, October, Julius & Mars,
Quatuor at reliqui ; dabit Idus quilibet octo.
Inde dies reliquos omnes dic esse Calendas.

All this was counted backward. We begin the year with the first day of January, as did Julius Cæsar; and which is nearly at the time that the sun enters the sign Capricorn.

XIII. The *Egyptian years of Nebuchadnezzar* are all of 365 days, and the twelve months each of 30 days, which making only 360 ; they added
five

five days to the end of each year, which they called the fupernumerary days. Nebuchadnezzar king of Babylon began to reign in the year of the world 3257, and, by the agreement of all chronologifts, 747 years before the common æra. The æra and year of Nebuchadnezzar fhould be clearly determined, in order to be made ufe of in drawing lights from the aftronomic obfervations of Ptolemy. The *year of the Moors* was much the fame with that of the Egyptians.

XIV. The *Perfians* had anciently the *Yezdegird year*, which agreed in all refpects with that of Nebuchadnezzar, except that it began on the 16th of July, and that of Nebuchadnezzar on the 26th of February, of the Julian year. The five days that were added they called *Mufteraka*: but, under the reign of the fultan Gelal, they changed their year, and adopted the fpace of the folar year; that is, 365 days, 5 hours, 49 min. 15″, 0‴, 48⁗. They ftill reckoned 30 days to each month, and the 5 Mufteraka at the end of the year: but after inferting fix or feven times in the fourth year an intercalary day, they made once, in five years only, a biffextile. They called it the *Gelalian year*; and it proves that the Perfians have been, for time immemorial, very expert in aftronomy; that they knew very accurately the fpace of the folar year, and how to intercalate the days in the moft proper manner, in order to make the equinoxes and folftices fall always on the fame days of the year.

The

XV. The *Syriac year* agrees in all things with the Julian, except that the months bear other names, and that the beginning of this year falls in the month of October of the Julian year. Ulugh Beigh, Albateignius, and other oriental authors, count by Syriac years.

XVI. The *Attic year* of the Greeks is a lunar year, and consists of 12 months, which have alternately 29 and 30 days. But to prevent it from beginning at all the seasons of the solar year, the Greeks made a bissextile of 13 months, and counted the sixth month twice. So that in a revolution of 19 years, the 3, 5, 8, 11, 14, 16 and 19th, were always Bissextile years. The beginning of this year was fixed to the day of the new moon which immediately preceded the summer solstice. In the time of Meton and Eudoxus, they placed it on the 8th of June; and, in the time of Timocharis and Hipparchus, it was fixed on the 27th of July. The Greeks were of all people the most wretched astronomers, and their chronology is consequently full of confusion. The lunar year of the Macedonians agreed with the Attic, and the solar year with the Julian. The Macedonians sometimes divided the year, moreover, into four equal parts, on the sun's entrance into the four cardinal points, and they alloted to each quarter 91 days.

XVII. The *Arabic* or *Mahometan* year is a lunar year that has 354 days. But as the Arabs adopted

adopted the lunar aftronomic year of 354 days, 8 hours, 48 minutes, they fometimes inferted a day at the end of the year, fo that in the fpace of 29 years, the 2, 5, 7, 10, 13, 15, 18, 21, 24, 26, and 29th years were Biffextiles. Their months were alternately of 29 and 30 days; and in the biffextile years the laft month, Dulheggia, was alfo of 30 days. The firft year of this period began on the 15th July of the Julian calendar.

XVIII. The *year of the modern Jews* is alfo a lunar year of 354 days, and has twelve months that confift alternately of 29 and 30 days. They fometimes added to the month *Odar*, or March, another entire month of 30 days, which they called *Veodar*, or more than March. Their intercalary years are, in 19 years, the 3, 6, 8, 11, 17, and 19th. The Jewifh year begins on the day of that new moon, which, according to the moon's mean motion, is neareft to the autumnal equinox. Sometimes they refcind from the common year, as well as from the biffextile, a day of the month Kiftow, or December; fo that the common year then confifts of 353 days only, and the biffextile of 383. Sometimes alfo they add a day to each of thefe forts of years, and then the former is of 355, and the latter of 385 days; the reafon of which is, becaufe they muft not celebrate the new moon of the month Tifchri or October, on the 1, 4, or 6th days of the week, or begin the new year on thofe days,

as

as that would be contrary to the institutions of their ancestors.

XIX. The solar year of the Jews is exactly the same as the Julian. It is divided into four equal parts; which are called Tekuphas, and are severally named Tekupham Tischri, Tebeth, Nisan and Tamuz: and are distinguished by the sun's entrance into the four cardinal points, Aries, Cancer, Libra, and Capricorn; and these days they celebrate with great solemnity.

XX. The point of time, from whence any number of years is begun to be counted, is called a *period, era,* or *epoch.* The word era comes from the Latin *æs,* because the Romans marked their years with a kind of small brass nails. The difference between the terms era and epoch is, that the eras are certain points fixed by some people or nation, and the epochs are points fixed by chronologists and historians. The idea of an era comprehends also a certain succession of years, proceeding from a fixed point of time, and the epoch is that point itself. Thus the Christian era began at the epoch of the birth of Jesus Christ.

XXI. *Chronological characters* are those marks by which one point of time is distinguished from another; which, by its resemblance, might otherwise be mistaken for it. Now, as the eclipses of the sun and moon, the sun's entrance into the

Vol. III. D four

four cardinal points, the new and full moons, the relative pofitions of the planets, and other celeftial phenomena, can be calculated to the greateft precifion, they may be regarded as infallible marks of time. Therefore, when we know the year of any people, and find a fact related by an author according to the chronologic date of another people, and that author alfo makes mention of another event that happened at the fame time among the former people, we may find, by the known year of one of thefe people, the unknown year of the other. According to thefe two methods of calculating, we may alfo find, by years that are known, how many years have paffed between them and the time any event has happened, the precife date of which has not been marked by hiftorians. For example, the year that a prince came to the crown may not be mentioned in the annals, but we may find that in a certain known year of his reign there was a remarkable eclipfe of the fun; from whence we may eafily calculate the precife year that he began to reign.

XXII. Mathematic chronology teaches us, moreover, the method of reducing, by means of calculation, the different years and periods of different people to one common meafure; to compare the one with the other, and thus to find the precife time in which every event recorded in hiftory has arrived. By thefe means we are enabled not only to range the facts of various nations,

nations, whose history is known to us, with their dates, in a regular series; but also to reduce all these events either to the Christian era, or that of the creation of the world. To facilitate this business, the celebrated Joseph Scaliger has contrived a particular method, which we shall presently explain.

XXIII. The *cycle of the sun* is a revolution of years, at the end of which, the letters that mark the Sundays and other feasts return in the same order in which they were in a former year. This revolution is performed in 28 years. The sun has no particular relation to this period, and it is only so called because the letter of Sunday is principally sought after. Chronology furnishes rules also for finding the Sunday or *Dominical letter*, and consequently those of the other days of the week.

XXIV. The *cycle of the moon* is a revolution of 19 years; at the end of which, the new and full moon fall on the same day of the Julian year. This method was invented by Meton the Athenian, who first observed, that after this term the lunations were the same. But this lunar cycle will not hold true for longer than 310 years in succession. The number that shows the year when the lunar cycle begins is called *the golden number*.

XXV. The *epacts* are the supernumerary days and hours that the Julian and Gregorian months have more than the lunar months. These latter months being of 29 days, 12 hours, 44 minutes, 3 seconds, it follows that a common month of 31 days must have 1 day, 11 hours, 15 minutes, 57 seconds, and a month of 30 days will have 11 hours, 15 minutes, 57 seconds, more than a lunar month. The *annual epacts* form in like manner the difference between a solar or civil year, and a lunar astronomic year.

XXVI. The *cycle of indiction*, or Roman cycle, is a revolution of 15 years. This method of computing was made use of by the ancient Romans, and it is still used in bulls and apostolic rescripts, as well as in instruments drawn up by German notaries. It is not certain by whom, or for what purpose, this cycle was first invented; but, by comparing it with the number of years from the birth of Christ, its first year falls three years before our Saviour's birth; though it does not clearly appear that the indiction was then in use.

XXVII. The *Julian period* is a space of time that includes 7980 years. Scaliger, who invented this period, composed it of the solar cycle of 28 years, the lunar cycle of 19 years, and the indiction of 15 years. For these three numbers, multiplied into each other, produce 7980. If we suppose, therefore, that the world has not yet existed

existed 6000 years, this imaginary period goes higher than the creation. But as all the years since the creation bear distinct characters in all the three revolutions we have mentioned, Scaliger made good use of it to compare and reduce, with more facility, the years and epochs of different nations of the earth.

XXVIII. Modern Christians count the years from the birth of Christ; but the first Christians reckoned from Dioclesian, and which they called the Dioclesian era, or the year of martyrs. The Moors still make use of it in calculating their festivals, and call them the years of grace. We shall presently speak more fully of these different eras, and especially that of modern Christians.

XXIX. In the Christian calendar the *feasts* or festivals are divided into *moveable* and *immoveable*. The moveable feasts, or those that do not always fall on the same day of the year, are Ash-Wednesday, Good-Friday, Easter-Sunday, Ascensionday, Whitsunday, Trinity-Sunday, &c. The immoveable feasts are New-year's day, the Epiphany, Lady-day, St. John Baptist, Michaelmas, Christmas-day, &c. By virtue of the canons or decrees of the Council of Nice, "*The feast of Easter is to be for ever celebrated on the first Sunday that follows the first full moon after the vernal equinox; and if that full moon fall on a Sunday, Easter-day shall be kept the Sunday following.*" Mathematical chronology shews different methods of calculating,

culating, according to this decree, which is followed by all Christian nations, the day of the year on which Easter will always fall; as well in the Gregorian as Julian calendar.

XXX. Lastly, this part of chronology teaches the method of constructing a complete calendar, as follows: 1. To find the feast of Easter, and the dominical letter. 2. To divide the calendar into weeks, and regulate the moveable feasts by that of Easter; inserting at the same time the immoveable feasts, with the names of those saints that are appointed for each day. 3. To extract, from those tables that are called Ephimeres, the place of the sun and moon in the zodiac, as well as of the other planets; to find the rising and setting of the two former, the duration of the twilight, and the length of the days and nights; and to insert all these matters in their proper places. 4. To remark when a planet is visible to us, and when it is hid by the sun's rays. 5. At the beginning of each month to make observations on the seasons, and to give account of the eclipses of the sun and moon, and of other celestial phenomena.

XXXI. Thus far we have treated of mathematic chronology. We shall now, in as brief a manner as possible, make the analysis of historic chronology, or of that science which teaches to distinguish the several events related in history according to the order of time in which they happened.

pened. It is in this science that Julius Africanus, Eusebius of Cæsarea, George Cyncelle, John of Antioch, Denis, Petau, Cluvier, Calvisius, Usher, Simson, John Marsham, and many other learned men, have excelled. It consists of four principal parts, that form the foundations on which all its learned researches rest. These are,

1. Astronomic observations, and particularly on the eclipses of the sun and moon, combined with the calculations of mathematic chronology on the different eras and years of different nations.

2. The testimonies of credible authors.

3. Those epochs in history which are so determined and evident that no one has ever contested them.

4. Ancient medals, coins, monuments, and inscriptions.

We shall examine these four principal parts in the order they here stand, and conclude with some reflections on the uncertainty that still reigns, notwithstanding these lights, in chronological history.

XXXII. It is with great reason that the eclipses of the sun and moon, and the aspects of
the

the other planets, have been called public and celestial characters of the times, as their calculations afford chronologers infallible proofs of the precise epochs on which a great number of the most signal events in history have occurred. So that in chronological matters we cannot make any great progress, if we are ignorant of the use of astronomic tables, and the calculation of eclipses. The ancients regarded the latter as prognostics of the fall of empires, of the loss of battles, of the death of monarchs, &c. And it is to this superstition, to this wretched ignorance, that we happily owe the vast labour that historians have taken to record so great a number of them. The most able chronologers have collected them with still greater labour. Calvisius, for example, founds his chronology on 144 eclipses of the sun, and 127 of the moon, that he says he had calculated. The grand conjunction of the two superior planets, Saturn and Jupiter, which, according to Kepler, occurs once in 800 years in the same point of the zodiac, and which has happened only eight times since the creation, (the last time in the month of December 1603) may also furnish chronology with incontestable proofs. The same may be said of the transit of Venus over the sun, which has been observed in our days, and all the other uncommon positions of the planets. But among these celestial and natural characters of times, there are are also some that are named *civil* or *artificial*,

and

and which, nevertheless, depend on astronomic calculation.

XXXIII. Such are the solar and lunar cycles, the Roman indiction, the feast of Easter, the bissextile year, the jubilees, the sabbatic years, the combats and Olympic games of the Greeks, and hegira of the Mahometans, &c. And to these may be added the periods, eras, epochs, and years of different nations, ancient and modern. We shall only remark on this occasion, that the period or era of the Jews commences with the creation of the world; that of the ancient Romans with the foundation of the city of Rome; that of the Greeks at the establishment of the Olympic games; that of Nebuchadnezzar, with the advancement of the first king of Babylon to the throne; the Yezdegerdic years, with the last king of the Persians of that name; the hegira of the Turks with the flight of Mahomet from Mecca to Medina, &c. The year of the birth of Christ was the 4713th year of the Julian period, according to the common method of reckoning. Chronology teaches us to calculate the precise year of the Julian period on which all these epochs happened.

XXXIV. The testimony of authors is the second principal part of historic chronology. Tho' no man whatever has a right to pretend to infallibility, or to be regarded as a sacred oracle, it would, however, be making a very unjust judgment

ment of mankind, to treat them all as dupes or impostors; and it would be an injury offered to public integrity, were we to doubt the veracity of authors universally esteemed, and of facts that are in themselves right worthy of belief. It would be even a kind of infatuation to doubt that there have been such cities as Athens, Sparta, Rome, Carthage, &c. or that Xerxes reigned in Persia, and Augustus in Rome; whether Hannibal ever was in Italy; or that the emperor Constantine built Constantinople, &c. The unanimous testimony of the most respectable historians will not admit any doubt of these matters. When an historian is allowed to be completely able to judge of an event, and to have no intent of deceiving by his relation, his testimony is irrecusable. But to avoid the danger of adopting error for truth, and to be satisfied of a fact that appears doubtful in history, we may make use of the four following rules, as they are founded in reason.

1. We ought to pay a particular regard to the testimonies of those who wrote at the same time the events happened, and that have not been contradicted by any cotemporary author of known authority. Who can doubt, for example, of the truth of the facts related by admiral Anson, in the history of his voyage round the world? The admiral saw all the facts there mentioned with his own eyes, and published his book when two hundred companions of his voyage

age were still living in London, and could have contradicted him immediately, if he had given any false or exaggerated relations.

2. After the cotemporary authors, we should give more credit to those who lived near the time the events happened, than those who lived at a distance.

3. Those doubtful histories, which are related by authors that are but little known, can have no weight if they are at variance with reason, or established tradition.

4. We must distrust the truth of a history that is related by modern authors, when they do not agree among themselves in several circumstances, nor with ancient historians, who are to be regarded as original sources. We should especially doubt the truth of those brilliant portraits, that are drawn at pleasure by such as never knew the persons they are intended for, and even made several centuries after their decease.

XXXV. The most pure and most fruitful source of ancient history is doubtless to be found in the Holy Bible. Let us here for a moment cease to regard it as divine, and let us presume to consider it as a common history. Now, when we regard the writers of the books of the Old Testament, and consider them sometimes as authors,

authors, sometimes as ocular witnesses, and sometimes as respectable historians; whether we reflect on the simplicity of the narration, and the air of truth that is there constantly visible; or, when we consider the care that the people, the governments, and the learned men of all ages have taken to preserve the true text of the Bible; or that we have regard to the happy conformity of the chronology of the holy scriptures with that of prophane history; or if we observe the admirable harmony that is between these books and the most respectable historians, as Josephus and others; and lastly, when we consider that the books of the holy scripture furnish us alone with an accurate history of the world from the creation, through the line of patriarchs, judges, kings and princes of the Hebrews; and that we may, by its aid, form an almost entire series of events down to the birth of Christ, or the time of Augustus, which comprehends a space of about 4000 years, some small interruptions excepted, and which are easily supplied by profane history: when, we say, all these reflections are justly made, we must constantly allow that the scriptures form a book which merits the first rank among all the sources of ancient history. It has been objected, that this book contains contradictions; but the most able interpreters have reconciled these seeming contradictions. It has been said, that the chronology of the Hebrew text and the Vulgate do not agree with the chronology of the version of the Septuagint; but

but the foundest critics have shown that they may be made to agree. It has been observed, moreover, that the scriptures abound with miracles and prodigies; but they are miracles that have really happened: and what ancient history is there that is not filled with miracles and other marvellous events? And do we for that reject their authority? Cannot the true God be supposed to have performed those miracles which pagan historians have atttibuted to their false divinities? Must we pay no regard to the writings of Livy, because his history contains many fabulous relations?

XXXVI. The *epochs* form the third principal part of chronology. These are those fixed points in history that have never been contested, and of which there can, in fact, be no doubt. Chronologers fix on the events that are to serve as epochs, in a manner quite arbitrary; but this is of little consequence, provided the dates of these epochs agree, and that there is no contradiction in the facts themselves. When we come to treat expressly on history, we shall mention, in our progress, all the principal epochs. In order rightly to understand and to range each epoch in its proper place, it is necessary to remember the signification of the following terms, beside those we have already explained in the course of this chapter.

An

An *age* or *century* is the courfe of a hundred years, or folar revolutions.

Luftre is the fpace of five years. The poets make frequent ufe of this term.

Olympiad is a fpace of four years, which the Greeks counted from the celebration of one of the Olympic games to another. The firft Olympiad began in the year of the world 3228, and confequently 776 years before the common era.

Epoch: To what we have juft faid on this term, it is proper to remark here, that chronologers diftinguifh three forts of epochs: the firft they call facred; the fecond, ecclefiaftical; and the third, civil or political.

Era: Befide what we have faid in the twentieth fection, we muft here obferve, that the word probably took its rife from the ignorance of copyifts, who, finding in ancient manufcripts the letters A. E. R. A. *Annus Erat Regni Augufti*, made of them the fimple word era, or, as the Latins write it, *æra*.

The *Seleucian era*, from whence the Macedonians began to count, is alfo denoted by the *Grecian years*, of which the Jews principally made ufe after they were fubdued by the Macedonians. It began with the great Seleucus, furnamed

named Nicator, in the year of the world 3692, and 312 years before the vulgar era.

The *Spanish era* began with the year of the world 3966, and 38 years before the common era. This era is very famous in the councils, and in the ancient monuments, of Spain.

Anachronism is an error in the calculating or fixing of time. So Virgil committed an anachronism in making Æneas and Dido live at the same time, when there were 300 years distance between them.

Synchronism is, as we have already said, the concurrence of different events at the same time. A general synchronism is a description of all that happened in the various parts of the world at the same period.

XXXVII. Medals, monuments, and inscriptions, form the fourth and last principal part of chronology. It is scarce more than 150 years since close application has been made to the study of these, and we owe to the celebrated Spanheim the greatest obligations, for the progress that is made in this method: his excellent work, *De præstantia et usu numismatum antiquorum*, has shown the great advantages of it; and it is evident that these monuments are the most authentic witnesses that can be produced. It is by the aid of medals that M. Vaillant has composed

posed his judicious history of the kings of Syria, from the time of Alexander the Great to that of Pompey: they have been, moreover, of the greatest service in elucidating all ancient history, especially that of the Romans; and even sometimes that of the middle age. We shall have occasion to speak more fully of their use in the chapter where we expressly treat of medals and antiquities. What we here say of medals, is to be understood equally, in its full force, of ancient inscriptions, and of all other authentic monuments that have come down to us: as the famous *Arundel marbles*, which an English nobleman of that name purchased from the Turks in the Levant, by William Petre, whom he sent thither for that purpose. These marbles, which were ranged at London in the rooms and garden of the earl of Arundel, on the border of the Thames, were found in the island of Paros, and contain a chronicle, wherein the principal epochs of the history of the Athenians are exactly and distinctly marked, from the first year of the Cecrops, which began 1582 years before the Christian era. John Selden composed a book in 1629, the title of which is *Marmora Arundelliana*, wherein he explains these valuable antiquities. Who can say what happy discoveries of monuments, Fortune, propitious to letters, may have reserved for us in the ruins of Herculaneum: and which may serve as well to elucidate as to establish ancient history?

XXXVIII.

XXXVIII. Every reader, endowed with a juft difcernment, will readily allow that thefe four parts of chronology afford clear lights, and are excellent guides to conduct us through the thick darknefs of antiquity. That impartiality, however, which directs us to give a faithful relation of that which is true and falfe, of the certainty and uncertainty of all the fciences, obliges us here freely to confefs, that thefe guides are not infallible, nor the proofs that they afford mathematical demonftrations. In fact, with regard to hiftory in general, and ancient hiftory in particular, fomething muft be always left to conjecture and hiftoric faith. It would be an offence againft common probity, were we to fuffer ourfelves to pafs over in filence thofe objections which authors of the greateft reputation have made againft the certainty of chronology. We fhall extract them from their own works; and we hope that there is no magiftrate, theologian, or public profeffor in Europe, who would be mean enough to accufe us of a crime, for not unworthily difguifing the truth.

XXXIX. 1. The prodigious difference there is between the feptuagint Bible and the vulgate, in point of chronology, occafions an embarraffment, which is the more difficult to avoid, as we cannot pofitively fay on which fide the error lies. The Greek Bible counts, for example, from the creation of the world to the birth of Abraham, 1500 years more than the Hebrew and Latin Bibles,

Bibles, &c. 2. How difficult is it to ascertain the years of the judges of the Jewish nation, in the Bible? What darkness is spread over the succession of the kings of Judah and Israel? The calculation of time is there so inaccurate, that the scripture never marks if they are current or complete years. For we cannot suppose that a patriarch, judge, or king, lived exactly 60, 90, 100, or 969 years, without any odd months or days. 3. The different names that the Assyrians, Egyptians, Persians, and Greeks, have given to the same prince, have contributed not a little to embarrass all ancient chronology. Three or four princes have borne the name of Assuerus, though they had also other names. If we did not know that Nabucodonosor, Nabucodrosor, and Nabucolassar, were the same name, or the name of the same man, we should scarcely believe it. Sargon is Sennacherib; Ozias is Azarias; Sedecias is Mathanias; Joachas is also called Sellum; Asaraddon, which is pronounced indifferently Esarhaddon and Asarhaddon, is called Asenaphar by the Cuthæans; and by an oddity of which we do not know the origin, Sardanapalus is called by the Greeks Tenos Concoleros. 4. There remain to us but few monuments of the first monarchs of the world. Numberless books have been lost, and those which have come down to us are mutilated or altered by transcribers. The Greeks began to write very late. Herodotus, their first historian, was of a credulous disposition, and believed all

the

the fables that were related by the Egyptian priests. The Greeks were in general vain, partial, and held no nation in esteem but their own. The Romans were still more infatuated with notions of their own merit and grandeur: their historians were altogether as unjust as was their senate, toward other nations that were frequently far more respectable. And, with regard to the Jews in particular, it seems, whatever Josephus may say, that their nation, who possessed only that small country called Palestine, never made a sufficient figure in the world to attract the regard of the historians of other civilized people. 5. The eras, the years, the periods and epochs were not the same in each nation; and they, moreover, began at different seasons of the year. All this has thrown so much obscurity over chronology, that it appears to be beyond all human capacity totally to disperse it.

XL. Christianity itself had subsisted near 1200 years, before they knew precisely how many years had passed since the birth of our Saviour. They saw clearly that the vulgar era was defective, but it was a long time before they could comprehend that it required four whole years to make up the true period. Abbé Denis the Little, who, in the year 532, was the first among the Christians to form the era of that grand epoch, and to count the years from that time, in order to make their chronology altogether Christian, erred in his calculation, and led all Europe into his error.

They

They count 132 contrary opinions of different authors concerning the year in which the Meſſiah appeared on the earth. M. Vallemont names 64 of them, and all celebrated writers. Among all theſe authors, however, there is none that reckon more than 7000, nor leſs than 3700 years. But even this difference is enormous. The moſt moderate fix the birth of Chriſt in the 4000th year of the world. The reaſons, however, on which they found their opinion, appear to be ſufficiently arbitrary.

XLI. Be theſe matters, however, as they may, the wiſdom of Providence has ſo diſpoſed all things, that there remain ſufficient lights to enable us nearly to connect the ſeries of events: for in the firſt 3000 years of the world, where profane hiſtory is defective, we have the chronology of the Bible to direct us; and after that period, where we find more obſcurity in the chronology of the holy ſcriptures, we have, on the other hand, greater lights from profane authors. It is at this period that begins the time which Varro calls *hiſtoric:* as, ſince the time of the Olympiads, the truth of ſuch events as have happened ſhines clear in hiſtory. Chronology, therefore, draws its principal lights from hiſtory; and, in return, ſerves it as a guide: as we ſhall ſee in the following chapters.

CHAP.

CHAP. IV.

On HISTORY in general, and its divisions.

WE are now come to one of the most pleasing prospects in the vast empire of the sciences; to one of the most important objects of universal erudition; to a study worthy to engage the attention of the first of mankind. History is now the subject of our reflections. All who have hitherto treated on this interesting part of literature, and have attempted to point out the most proper method of attaining it, have constantly repeated what Cicero and their other predecessors, ancient and modern, have said in praise of history. We do not think it necessary here to enumerate those encomiums, but shall endeavour to add to their sagacious reflections some further remarks on the utility of this admirable science.

II. Ignorance was ever disgraceful to humanity; and it is more especially so in an age which offers

offers so many sources of instruction, that it cannot proceed but from negligence or idleness. Even among the least civilized people, history has been at all times held in esteem. Before the use of letters were known to mankind, they transmitted to their posterity the actions of their ancestors, their heroes, and the founders of their nations, by hymns or songs, in which poetry, ignorant as it then was, constantly mixed fable with truth. It is for this reason, doubtless, that the most ancient people, and even the Greeks, confounded these two terms, calling history sometimes fable, and fable, history. For the word *history* is derived from the Greek verb ιϛοϱεῖν, which signifies to contemplate or consider. Under this collective term, therefore, they comprehended not only the knowledge of things past, but also mythology, Esopean and Milesian fables, romances, tragedy, comedy, pantomimes, &c. But words like these, which are too universal, constantly discover the indigence of a language; for, by comprehending too many objects, they serve only to create confusion in our ideas, as well as in the sciences. It is for this reason that the most sagacious of modern literati endeavour to dispel the chaos of erudition, and to give to each word, each term of art, a fixed and determinate signification, and not to comprehend, under the denomination of a science, any objects that do not absolutely and necessarily relate thereto.

III. Ac-

III. According to reason, therefore, as well as the practice of modern writers, *history is a true relation of real facts and events that have occurred in the world.* If mere curiosity did not excite in the mind of man a rational desire of knowing what has passed on the earth, that is remarkable and interesting, from the creation to our own time; and if the knowledge of all these matters did not improve the understanding of those who are destined to live among the intelligent part of mankind, and did not render their conversation more pleasing, more striking and instructive, yet would they find, in the study of history, numberless other advantages, that are still more important, and that prove its excellence. History, being the faithful depository of all the actions, good and bad, of the whole race of mankind, who have lived in all ages, and have performed any distinguished part on the theatre of the world, forms the most powerful incentive to virtue, and preservative from vice. The most succesful usurper, the most absolute and cruel tyrant, would not have his memory appear loaded with infamy in the eyes of posterity. To cover the iniquity of his enterprises, he accompanies them with manifestoes, and other memoirs of justification. But history here tells him, that his efforts are vain, that the time will come when his iniquity will be unvailed, and the secret folds of his heart laid open; when neither the arts of his worthless ministers, nor the eulogies of venal pens, will be able to defend him: that posterity will

will be his judge; and that the only method of obtaining a favourable sentence, is, by performing worthy actions: that true glory is never to be found but in real merit: that history flatters not: that it treats the wicked even with an inexorable severity; and that it pays no respect to sceptres or diadems.

IV. History likewise forms, so to say, a course of experimental morality and politics, where the causes and effects of human actions are exposed to our sight. It is a scene where the characters and precepts of Theophrastus, la Bruyere, and Shaftesbury, are put in action. Here all takes a body, a mind, a soul. Experience, which costs mankind so much time, and so many errors, is here acquired at once, or, at least, by a single study. Princes especially, and they whom Providence has called to the government of a people, or to the dictating of laws, should never be ignorant of this science: for, though they ought not to draw their maxims of government, or their laws, from history itself, seeing that would render them pitiful imitators, by chance, of the wisdom, but much more frequently of the folly and depravity of past ages, history, nevertheless, will warn them of numberless rocks that are but just covered by the vast ocean of politics, and against which they would be in continual danger of rushing, if they were not directed by this skilful chart.

V. We

V. We have three objects to explain in this chapter; which are,

1. The manner of writing history:
2. The manner of studying it:
3. The different divisions, or species of history.

With regard to the manner of writing history, the first fault that we find in all historical writings, ancient and modern, and which appears to us of no small magnitude, is, that they consist of a mere description of those wars that have desolated the earth from the origin of the human race. It should seem as if mankind found nothing great in nature, nothing worthy their attention, but that which ought to cover them with shame and confusion: that which arises from their depravity, a mad desire of victory, of destroying each other; a barbarous custom of maintaining their pretensions by the force of arms; of imagining that superior force gives right; and the folly of placing a vain honour, a false glory, in their brutal quarrels and combats. Follies are frequently contagious: that of heroes has infected their historians: blood must be constantly spilt: if they were to place only one man upon the earth, they would make him fight, either against the gods or devils, or with serpents and monsters, or else with his own shadow, rather than paint him peaceful and amiable. If they should suppose two men to exist, it would be merely with a design that they might destroy each other, or at least that one of them might murder his companion.

nion. When they made Cadmus fow the earth with teeth, from whence men fprung up, it was neceffary that thefe firft of human race fhould immediately attack and butcher each other.

Barbarians! to whom no object appears great but that of war! The nurture of the human race, their eftablifhments, their migrations, the founding of cities and colonies, the progrefs of the human mind in the arts and fciences, grand inventions and difcoveries, as that of navigation and a new world, and a thoufand like objects; Are not thefe worthy of regard? A king came to the crown on fuch a day, in fuch a year: without the leaft reafon he attacked fuch a people, and after that fo many others; or he was himfelf attacked; and fuch were the confequences of his wars, he overthrew fo many cities, he took fo many prifoners, and left fo many dead upon the field; and at laft this mighty monarch himfelf is killed, or he dies with remorfe in his bed. You have here, in a few words, the fubftance of hiftory in general; fome little ornaments of moral and political reflections apart.

VI. The fecond fault of hiftorians is, the bad proportions they obferve in the arrangement of their works. Each hiftory, whether univerfal or particular, refembles a peacock, who, to a very fmall head, and a body indifferently large, has joined an enormous tail, which continually extends as it approaches the extremity. The beft best writers of history are faulty in this respect. Every one can repeat those excellent lines with which Tacitus begins his annals; and when they shall remark the concision he there observes, and compare it with the prodigious number of animadversions that are spread over his history, and the prolixity with which he concludes, they will be convinced that our observation is just. It is to be wished, therefore, that the writers of history would acquire the art of extending their introductions, and of contracting their conclusions, that there might be more uniformity in the parts, more regularity and harmony in the whole. Curious and learned researches, pleasing and useful reflections, are very natural amplifications. And why are not facts that occur in the beginning of a history as worthy of our attention as those of latter times? We know there are many who are of a contrary opinion, but we think they deceive themselves. All the details of recent events serve only to promote chicanery and the quarrels of sovereigns: their ministers make use of them to produce arguments in defence of their pretensions. But, should history be debased to such purposes as these? Are there not memoirs, periodical productions, and archives, sufficient to kindle these disputes, to furnish deductions, and to support these literary wars?

VII. All modern capital histories have likewise the fault of being highly prolix. What life is sufficiently long, what eyes are good enough,

enough, and what memory is strong enough, to read and retain these works? Those of de Thou, Mariana, Rapin Thoyras, Barre, Daniel, and the rest of this class? By naming a few historians only, it is easy to enumerate several hundred folio and quarto volumes: and if we reflect that M. le Long, in his Historical Bibliotheque, has produced the names of more than twenty thousand authors who have wrote the history of France only; and that the late count de Bunau collected above thirty thousand German historians, whom they call *Scriptores rerum Germanicarum,* we may easily conceive how enormous a chaos all this must form, and what indefatigable labour it would require to wade through this vast, barren desart of erudition. In proportion as the world increases in years, this historic body increases in bulk, and must at last sink by its own weight. All that can be done in this case is, to regard these voluminous works as historic dictionaries, that are not to be read, but consulted occasionally.

VIII. Independent of these faults, which the historian ought to avoid, there are also some precautions to be observed, in order to which it will be proper here to lay down certain precepts. 1. No one should attempt to write a history without a perfect knowledge of all its parts. By constantly running, a man may excel in the race, but he will never excel as a historian, merely by writing. It is true, that in the course of the

work

work he may frequently make curious and useful discoveries, but the ground of the subject on which he is to treat ought to be familiar to him; he should therefore well consider his strength before he attempts the enterprise. 2. When a choice is judiciously made, he should examine the sources (*fontes*) from which the facts are to be drawn. Original memoirs, manuscripts, archives, and other scarce papers, are of an inestimable value to an historian, by enabling him to present the public with subjects that are new and interesting. But, if he be not provided with these, he ought at least to consult the historical bibliotheques, in order to inform himself of those authors who have wrote on that part of history; to procure their writings; to make a careful examination of them, and to extract all that can be of use to his subject. A judgment more than common is here necessary, in order to distinguish the false, the fabulous, exaggeration and prejudice, from truth and impartiality; and to determine the degree of credibility that is to be assigned to each author. The chapter, in which we shall treat of the knowledge of authors, will contain some further instructions on this subject.

IX. When the historian is provided with these materials, he should, 3dly, begin his work by extracting those articles that are to compose his history. And here it is indispensably necessary to make a judicious choice, and to range them in a clear order. Nothing that is interesting should be

be omitted, and nothing which can be omitted should hold the place of that which is interesting. An historian should faithfully relate all that is commonly said of an event, and of its circumstances, without being obliged to be answerable for the strict truth of what they may sometimes contain that is marvellous or incredible. He that would write the history of Rome, and should pass in silence the tradition of Romulus and Remus being suckled by a wolf, would commit an egregious fault. No reasonable man can believe that Hatton, archbishop of Mayence, was devoured by rats, and yet it would be unpardonable to make no mention of such report, when writing the history of that archbishoprick. An able writer will endeavour, in the first place, to reconcile these sorts of popular traditions with the truth, and which if he cannot effect, there is a certain manner of relating such stories, by which the reader will immediately perceive that the historian gave them no credit. The following words of a celebrated author, contain also an important observation : " There are a thousand
" incidents that are interesting to a cotemporary,
" but which are lost to the eyes of posterity;
" and which, disappearing, leave those great
" events only visible, that have determined the
" fate of empires. Every thing that is done,
" does not deserve to be written." For the rest, he will produce a mere chaos only, painful and disgustful to the reader, who, after having made choice of the matters he would relate, does not

reduce

reduce them to a regular chronology, by making a rough draft of the history he proposes to write, by carefully observing the several epochs, by never losing sight of the synchronisms, and by taking special caution to avoid all anachronisms, which are the most unpardonable faults in history.

X. 4. Particular anecdotes are of the highest use in ornamenting a history, but we should take care not to be too lavish in these ornaments, for, by that mean, they become insipid. The historian should therefore be moderate in the use of these, and have constantly before his eyes the gravity and majesty of history. 5. We have so often said that an historian should be impartial, that he should have neither country, nor particular religion, and the observation is itself so manifest, that it may seem almost superfluous in this place. An excessive predilection, notwithstanding, is a fault with which the generality of French historians may be justly reproached. They see nothing great, but what is to be found among themselves. They are so much possessed with this prejudice, that, in an universal history, they fix the periods by the annals of their own monarchy, and make, for example, an epoch of the time that Lewis XIV. after the death of his prime minister, resolved to govern by himself. We should be glad to know of what importance this was to the rest of the world. It appears to us to be a mean and ridiculous piece of flattery.

XI. The

XI. The style is so important an object in writing a history, that we cannot sufficiently recommend an attention to it. How excellent soever are the matters that a book contains, is of little importance, if, for want of perspicuity and elegance in the writing, we cannot be induced to read it. If, in the choice of a style, we were obliged to make use of that which is very concise or very diffused, we should incline to the former. The point of perfection is, however, in a just medium. Style is a gift which every writer receives from nature. We know of no two that are precisely the same. If we may be permitted to propose the best French models of style, we think they may be found in the History of Charles XII. and in the Age of Lewis XIV. by M. Voltaire; in the Revolutions of the abbé Vertot, in the Historic Pieces of the abbé St. Real, in the Universal History of M. Hardion, and in some other modern historians. The style that M. Bossuet, bishop of Meaux, has employed in his Discourse on Universal History, is inimitable, and might serve as a capital model, if that prelate had not endeavoured after too much eloquence, and if he did not sometimes do violence to the truth, in order to be always favourable to religion; of which he appears to be the panegyrist.

XII. Facts and events make the body of a history; the instructions they afford make the soul of it. A history must resemble a journal or gazette, if the author does not introduce those
efficacious

efficacious reflections, which sometimes discover the secret causes of human actions, and sometimes point out their consequences. And here a bold and lively genius is necessary; one that can break through those obstacles which stop the vulgar mind, and that can produce thoughts where truth and novelty are united: it is here that an uncommon discernment is requisite; a marvellous sagacity that can penetrate the human heart, that can make its way into the cabinets of princes, and into the minds of ministers and generals; that can unfold what passes there, and that judges of their thoughts by their actions, rather than by their words and writings. All these reflections, moreover, should arise from the subjects themselves, and not be forced into the work. They should likewise be made with moderation, and not in the manner of Tacitus, who, so to say, drowns all events in the sea of politics. Lastly, as all the reflections that a history contains should tend to form the heart as well as the mind of the reader, to render virtue amiable, and meliorate the human race; all malevolent satire, all fallacious reasoning, all impiety, all ridicule of religion, are at once ill placed, and highly blameable in history. The writer who shall think to shine by these means, will find he makes a very different appearance in the eyes of the sagacious part of mankind, though he may sometimes dazzle the ignorant: and he will be the less esteemed for these railleries, as they are far from being so difficult to produce as some may imagine.

XIII. It is a general custom to make from a history a gallery of portraits, formed of the characters of the principal actors that are introduced on the scene: to paint their exterior figures, as well as their manners, passions, &c. We do not entirely disapprove of this custom; but whoever shall consider how difficult it is for a painter to catch the likeness of an object that he has before his eyes, and of a discerning person to paint the mind even of those with whom he is intimately acquainted, will easily judge what kind of regard is to be paid to these sort of portraits that are drawn several ages after the existence of their originals; the features of which are collected from ancient authors, who frequently knew no more about them than the modern painter. One of the best drawn portraits we have ever read, is that which M. Duclos has placed at the end of his excellent history of Lewis XI. And yet we imagine, that if any courtier who was admitted to a familiar acquaintance with that monarch, was to come now upon the earth, he would scarce know his master. As to those formal panegyrics which some historians make on their heroes, there is nothing which appears to us more insipid, and more unworthy of the truth and gravity of history.

XIV.

XIV. 9. Almoſt all the ancient hiſtorians have an idle method of crowding their hiſtories with a number of harangues. We will for once aſſume a deciſive tone, and pronounce all theſe harangues, that are pretended to have been addreſſed to whole armies, to be either fictions or abſurdities; for it is impoſſible for the commander of an army to make himſelf heard, even by a whole regiment that is neareſt to him, and ſtill much leſs by a numerous army extended by ranks and files. For even the proper officers would ſcarce be able to make the words of command heard on the day of action, though pronounced in monoſyllables, and with a loud voice, if the ſoldiers were not previouſly acquainted with them. The general therefore, who ſhould ſtrain his throat with making a long florid harangue before a battle, to an army that could not poſſibly hear it, would be deſervedly regarded as a madman. Thoſe orations which are ſuppoſed to be made from the roſtrum to an aſſembly of the people, by an ambaſſador to a monarch, or by a public orator to a ſenate or council, are more juſt, more natural and probable. But even ſuppoſing them to be true, they ought not to be very frequently introduced: for they are a ſort of machinery that loſe their power when too often uſed.

XV. 10. Laſtly, in writing a hiſtory we may ſometimes make an advantageous uſe of letters, diſcourſes, reflections, ſayings and writings

tings of those kings, heroes or magistrates of whom we are speaking, by relating them either entire or in abstract: and this is an advantage that ought not to be neglected; for nothing gives history a greater air of veracity, or better proves its authenticity. When with these precautions the writer is sparing in his accounts of wars, when he avoids all long descriptions of battles and sieges, which, after all that can be said, from the time of Joshua and Cyrus down to the present age, strongly resemble each other, and are attended with a disgustful uniformity; and if instead of these he explain the causes of grand revolutions and remarkable events, and especially if he be strictly true, judicious and impartial in his relation, he may safely indulge in the pleasing reflection of having wrote a history worthy the approbation of the present age, and of posterity.

XVI. Most of the precepts we have here given for the manner of *writing* history, have an intimate connexion with the manner of *studying* it. Whoever would apply to this study, ought in the first place to recollect all that we have said in the preceding chapter on chronology: for if we do not carefully distinguish the several eras, periods and epochs, we shall never be able to form in our minds a regular and fundamental system of history, and to range each fact in its proper place. The method that appears to us of all others the most eligible, is nearly contained in the following particulars.

XVII. We would begin by placing before the eyes of our pupil a sketch, the mere outlines of universal history, or chronological tables; or rather a large historical and chronological chart, such as that of which Justus Lipsius conceived the idea, and which we have frequently intended to execute, had not other very different occupations diverted our attention. When we perceived that this general draught had made a sufficient impression on the mind of our pupil, we would make him read aloud the most concise and finished abridgment of history we could procure; taking particular care to remark to him, as he went on, the several synchronisms or events that happened at the same period among the different nations of the earth. By this mean we should by degrees fill up our sketch, and provide our pupil with what is called the thread of history. This preliminary study would take up but little of his time, and would be of great use to him during the whole course of his life. We have elsewhere wished, that the histories of all nations, ancient and modern were wrote on the model of the chronological abridgment of France by the president Henault, which we cannot too often repeat; and we have the high satisfaction to see that our wish is daily carrying into execution.

XVIII.

XVIII. We would then pass with our pupil through a cursory lection of those authors, as well ancient as modern, that are called the sources of history (fontes): of these we would choose but a small number, and would take particular care to select those only whose authenticity appears unquestionable. After this, we would go through a complete course of universal history, which we would endeavour to enliven with moral, political and military reflections, with critical remarks on dubious facts, &c. And here especially, we would place before his sight the portraits of those great men who have filled the throne, or directed the cabinet, have commanded armies, adorned the mitre, or illumined the sciences. We would endeavour here to point out their virtues and their vices, their sagacious and their futile actions, their glory and their shame. We would paint the tyrant, the rapacious minister, the senseless or brutal commander, the bigoted priest, and the idly laborious scholar, in their proper and disgustful colours: in a word, it is here that we would endeavour to draw all that comprehensive and and lasting utility which history is capable of affording.

XIX. In the last place; during the remainder of those years which are consecrated to his education, we would teach him the history of each particular modern nation, beginning with that of his own country: and here we would point out
the

the sources from whence he might draw the history of each particular province or district, the annals of each city, &c. And in the course of our progress we would study ecclesiastic history, that of litterature, and those other matters, with the enumeration of which we shall conclude this chapter, and which will make the subjects of some of those that follow.

XX. History in general is divided into
1. Civil or political history, which relates all the revolutions and all the memorable events that have occurred in governments; and gives an account of the method by which all nations have been founded, established, maintained and improved; of their increase, decline, and final dissolution.
2. Military history, which recounts the wars that each people have sustained, their battles and sieges, the good and bad success of all their military operations; those generals that have distinguished themselves, &c. Xenophon, Polybius, Vegetius, Quincy, and many others, have wrote military histories.

History, as well civil as military, is subdivided into
The ancient.
That of the middle age.
The modern.
The three following chapters will explain these subdivisions, and give their analysis.
3. Re-

3. Religious hiſtory; which treats in general of the religion and worſhip of all nations, both ancient and modern; of religious ceremonies, and of the origin, progreſs and decline of each religion.

4. Eccleſiaſtic hiſtory, or that of the Chriſtian church in particular; which teaches the origin and revolutions of the true religion, of the oppoſitions and perſecutions it has ſuſtained, of the ſucceſs it has met with, and of the triumph it has finally obtained; from the commencement of the world to the preſent time. It comprehends alſo the hiſtory of the various hereſies and ſchiſms of the popes and reformers, &c. and is ſubdivided into

The hiſtory of the church of God under the Old Teſtament. And

The like hiſtory under the New Teſtament.

5. The hiſtory of litterature, which treats of the progreſs of the human mind in general, and comprehends

Phyſical or natural hiſtory, which relates all that has arrived, or rather all that has been diſcovered and obſerved, that is remarkable, from the time of the creation; either in the heavens, in the elements, or among men, animals, inſects, plants, and in general among all the parts and productions of nature.

Philoſophic hiſtory, that teaches the progreſs of philoſophy among all the people of the earth.

The

HISTORY. 89

The history of erudition, which gives an account of the state of the other sciences among all nations.

Technical history, that treats of the progress of the arts, as well liberal as useful.

6. The history of the learned; which relates the lives and productions of the learned men of all ages, in those works that are called Biographies.

7. Miscellaneous history (Historia mixta vel miscellanea); which contains all sorts of anecdotes, political, ecclesiastic, military, literary and civil, that are of any importance, and that are not included in pragmatic or political history.

XXI. They make in the schools still other divisions of history, as into

Sacred and profane.

Universal or fundamental, and particular or special.

Real and poetical or fabulous.

Antediluvian and postdiluvian.

European, Asiatic, African, American, &c. &c.

But without attending to these divisions, which are founded less in the nature of the objects that relate to history, than in the imaginations of those who profess it, and which, far from elucidating this science, serve only to perplex it, by overloading the memory; we shall content ourselves with thus merely enumerating the

principal of thefe divifions, that our readers may not be quite ignorant of them, and fhall immediately pafs to the analyfis of the real objects of hiftory.

CHAP. V.

ANCIENT HISTORY.

WE can write that only which we know, and in all the hiftoric fciences, we can learn that only which is written. From this inconteftable axiom we may draw fome inftructive confequences. The firft is, that our ancient hiftory cannot go higher than Adam, who is reprefented to us by Mofes (the moft ancient of all thofe authors and hiftorians whofe works have come down to us) as the origin of the human race. We know indeed, that in working a quarry of porphyry they have lately found, in the middle of a block of a prodigious fize, a bar of wrought iron, and that according to
the

the calculations of the moſt ſkilful naturaliſts, it would require more than ten thouſand years for ſo large a maſs of that hard marble to grow round a bar, and if they knew the art of forging iron more than ten thouſand years ſince, the world muſt be much older than Moſes makes it to be. We know alſo that the world has numberleſs other natural marks which ſeem to prove an antiquity ſtill far greater. We are not ignorant moreover of all the arguments that may be drawn from the chronology of the Chaldeans, Egyptians, and Chineſe, which go vaſtly higher than that of Moſes: but it ſeems to us, at the ſame time, that the world alſo affords numberleſs marks of a recent ſtate, which counterbalance the former, and at leaſt reduce the ſeveral arguments to conjectures only. All the chronologies of the Chaldeans, Egyptians and Chineſe, are founded moreover entirely on traditions, and on certain vouchers that are equally equivocal and ſuſpicious. During the firſt ages of all nations the art of writing was unknown. It was a long time before letters were invented: and what confidence can be placed on a chronology, ſupported only by traditions, and, what is worſe, by the traditions of the Orientals, whoſe heated imaginations have at all times produced ſwarms of reveries, fables and extravagancies?

II. But let us ſuppoſe for a moment that there have been Preadamites. This might injure

jure us much as Christians, because if this fact could be established, it would render the Mosaic history very doubtful; but it would be of very little use to us as historians. For what could history have to do with these preadamite people, of whom we know nothing, either by writing or tradition? Beside, all the ancient chronology of the Egyptians and Chinese is the most wretched that can be conceived, built on the weakest foundations, and so confused, that it is impossible to deduce any one fact from it that bears the least character of authenticity. Reason and religion therefore equally require that we begin our ancient history with the creation of the world, according to the account of Moses, and consequently that we regard Adam as the first of mankind.

III. The second consequence we draw from our first principle is, that the greatest part of those ancient people, who inhabited the different countries of the earth, being ignorant of letters, could not transmit the history of their own nation, even to their descendents, and still much less to others. There may have been thousands of nations, whose very names are not come down to us. Some of these names indeed were by chance transmitted by oral tradition to those people who first knew the use of letters, and particularly to the Greeks: but these Greeks were at once credulous and fallacious. Herodotus, the first of their historians, readily believed all the fables
and

and traditions which the Egyptian prieſt had told him on his travels; and of theſe he compoſed nine poems in proſe, each of which he dedicated to one of the muſes, and recited them one after the other at the Olympic Games, and which the people greedily received, admiring all his marvellous ſtories.

IV. The third conſequence we draw from our principle is, that ancient hiſtory is leſs the knowledge of what has really happened in the world, than of that which hiſtorians have related, and what they have affirmed as facts. And, in truth, is not this ſufficient? Does it not contain ſufficient matter fully to ſatisfy our curioſity? Do we not find in hiſtory, as it now is, ſo vaſt a compilation of facts and events, that the longeſt life, and moſt happy memory, is ſcarce ſufficient to learn and retain them. Is not the time of antiquaries, critics, and commentators, fully employed in learned reſearches? And of what conſequence is it to us, after all, to know the exact truth of each ancient fact or event? Would this preciſe knowledge render us in any reſpect better, or can it in the leaſt contribute to our happineſs? On the contrary, it is eaſy to prove, that the preſent generation are more obliged to an ancient hiſtorian who has recounted an event ſomewhat fabulous in its circumſtances, but in a manner that is intereſting, agreeable and uſeful, than to one who has related facts that are preciſely true, but in a manner cold, dry, and diſintereſting.

interesting. A fable teeming with instruction appears to be, in this case, far preferable to a barren truth.

V. We by no means despise the efforts of those men of transcendent genius and indefatigable application, who pass their whole lives in making judicious inferences, or ingenious conjectures, in order to reconcile passages, discover truths, or diffuse lights over the history of the first ages of the world: but we think, at the same time, that their labours are not accompanied with any real certainty, or any direct utility to mankind. While I was writing the above I discovered, from the window of my closet, a large hole in my garden wall; I enquired among my domestics, I consulted even my chaplain, concerning the cause of this hole. Each of them assigns the reason at a venture, and all of them support their opinions with warmth. An arch fellow steps up and tells us we are all in the wrong, discovers the real fact, and leaves us all sufficiently confounded. I imagine the inquirers into the facts of ancient history are frequently in the same circumstance with me and my wall.

VI. When we duly consider the matter, we find that ancient history may be divided into two parts. The first contains the history of the Jews, or Hebrews, or of those who are called the people of God. Independent of that religious faith which this history requires of Christians, seeing

seeing it forms the basis of their religion, it merits likewise a peculiar regard by all mankind, considering it merely as profane annals, 1. Because it carries with it the marks of veracity, while the ancient history of other nations, especially during the first ages, is manifestly nothing more than a collection of fables. 2. Because it contains a chronological succession of events, almost without interruption, which we do not find in any other history; as we shall see further on. And 3. Because it forms a general scale, a common measure of chronology for all other histories; for, without this, we should not find in any of them any measure of time, nor any certain epoch; all ancient history would be a mere chaos, impossible to be reduced into any form: a region covered with impenetrable darkness.

VII. This history, which, on more than one account, deserves the title of sacred, admits of many divisions, of which we shall here mention two only, and these appear to us natural, and remarkable by the importance of their epochs. For, in the first place, we may consider the Jews under four kinds of governments; as,

1. The patriarchal, under 22 patriarchs.
2. The judiciary, under 22 judges.
3. The royal, under 22 kings.
4. The sacerdotal, under 22 pontiffs; among whom some have born the title of kings, as Aristobulus, Alexander, Hircan, Antipater, Herod, &c.

The

96   UNIVERSAL ERUDITION.

The Jewish history differs from all others in this particular division. It may, moreover, be divided into different ages, which may be thus fixed:

	Years.
The 1. Beginning with the creation of the world, and ending with the universal deluge, comprehends	1657
2. Beginning immediately after the deluge, and ending with Abraham, in the year of the world 2083, comprehends	426
3. Beginning with Abraham, considered as the father of the Jewish nation, and ending with the departure from Egypt, which was in the year 2513, comprehends	430
4. Begins with the going out of Egypt; when Moses, becoming the legislator and judge of the people of Israel, conducted them through the desart, and left to Joshua, his successor, the care of the conquest of the country of Canaan, and the establishment of the Jews in that promised land. This age begins with the Judaic republic, and continues to the time of the establishment of the royalty: it comprehends	396

5. Begins with the reign of Saul, the first king of the Jews, who was anointed by Samuel in the year 2909, and concludes with the end

of the captivity of that people in Babylon when Cyrus permitted them to return, in the year of the world 3468. This period includes also the divifion of the Jewifh monarchy, with the eftablifhment of the kingdom of Judah and that of Ifrael; it confifts of - - - - 559   *Years.*

6. Begins with the liberty that Cyrus granted to the Jews, and ends with the birth of Jefus Chrift, which was about the year of the world 4000, and confequently comprehends - 532

In all, - 4000

This epoch includes, among the reft, the wars that the Jews had to fuftain againft the Romans, and which ended in rendering them tributary to that monarchy.

VIII. At the beginning of the feventh age, there appeared, among the chofen people, *the Meffiah, the Saviour, the Redeemer of mankind.* Forty years after the death of Chrift, Jerufalem was deftroyed by Titus, the fon of Vefpafian, and, after him, emperor. The Temple was pillaged, the inhabitants partly exterminated, and partly carried away captive, and difperfed over the face of the earth. Thus finifhed the republic of the Jews; who, from that fatal period, have never been able to affemble as a nation. They who followed the Meffiah and embraced his holy doctrine,

doctrine, which may be said to be grafted on that of the Hebrews, called themselves, after his name, Christians; and dated the epoch of their history from the birth of Christ. This epoch, as we have said, began about the year of the world 4000; and, to the time of writing this work, it has continued 1765 years. So that, without entering into a minute chronology, the world, according to common opinion, has subsisted, from the epoch of the creation to the present time, 5765 solar years of 365 days.

IX. The history of the Jews, as we find it in the holy scriptures, and as it is confirmed by Josephus, one of the best historians the world has produced, serves also to diffuse great lights over the histories of those ancient people with which this first nation had wars, alliances, or connections: and in these histories, fable is consequently less mixed with truth, than in those of other ancient nations, which are founded entirely upon doubtful traditions and monuments. We are, however, to draw, from profane authors also, all information that can be of any use in elucidating the histories of these nations, and of reducing them into the form of a system, however imperfect it may be. But, notwithstanding these aids, and all the pains that have been taken, there are still many chasms to be supplied in these histories.

X. Of

X. Of all those ancient nations, whose names, as well as their actions, have not been destroyed by the length of time, the distance of place, and the ignorance of letters, there now remain only,

1. The history of the Moabites, from their founder, Moab, the son of Lot, to the time of Nebuchadnezzar.

2. The history of the Ammonites, from Ammon to the same Nebuchadnezzar.

3. The history of the Midianites, from Midian, the fourth son of Abraham, to their two last kings, Zeba and Zalmuna, who were vanquished by Gideon.

4. The history of the Edomites, the descendants of Edom, the son of Isaac, to the time of Joram, the king of the Jews, by whom they were destroyed.

5. The history of the Amalekites, whose founder was Amalek, the grandson of Esau, to the time of Saul and David, when they no longer subsisted as a nation.

6. The history of the Canaanites, properly so called, from their founder Canaan, the son of Ham, to the time of Solomon, when they were confounded in the common name of Phœnicians

7. The history of the Philistines, from Mizraim, the son of Ham, their founder, to the time they were in part overcome by king Hezekiah, when their capital, Asdod, was destroyed by the Assyrians; and finally, to the time that

the last kings of Gaza, of that nation, were vanquished by the Egyptians, and their nation entirely extirpated.

8. The history of the ancient Syrians, as well those of Zobach, as those of Damascus, from Rehob, the first king, who lived in the time of David, to the reign of Jeroboam, who destroyed Damascus.

9. The history of the Phœnicians, from Agenor, the first king of Sidon, who reigned a short time before the Trojan war (though, according to Josephus, Sidon, the eldest son of Canaan, gave his name to that city and the country round about) to the time that Sidon, as well as Tyre, were reduced under the yoke of Alexander the Great.

10. The history of the Assyrians, from Pul, or Phul, to Sardanapalus. The capital of this empire was Nineve.

11. The history of the Babylonians or Chaldeans. This nation was more ancient than that of the Assyrians. Their founder was Nimrod, and Nebonassar their first king, whose consort was the famous Semiramis. Nebuchadnezzar, 17th king of Babylon, destroyed the kingdom of the Assyrians; and that of Babylon fell in its turn, in the reign of its twentieth king, Nabonadus (who was the Assuerus of the scripture) into the hands of the Medes and Persians.

12. The history of the Medes, whose empire arose out of the ruins of that of Assyria, or rather became formidable, when they were freed

from

from the yoke of the Assyrians. Their first king was Arbaces. The epoch of their grandeur was in the reign of their seventh king, Cyaxares, who conquered, in conjunction with Nebuchadnezzar, the city of Nineve. Assisted by the Persians, they also took, during the same king's reign, the city of Babylon; and lastly, Astyages (the Balthazar of the prophet Daniel) became possessed of the whole empire.

XI. 13. The history of the Persians, or Elamites; who owed their origin to Elam, the son of Sem. The first king, of whom there is mention made in the scripture, was Kedorlaomer. Cyrus, the founder of the new empire of the Persians, made himself master at the same time of those of the Medes and Babylonians. Their last king, Darius, surnamed Codomanus, was vanquished by Alexander.

14. The history of the Scythians, or Chomereans, who were also called Cimbri, or Celts, and were descended from Gomar, the eldest son of Japhet. Their first king was Scythes, a pretended son of Hercules; and their last Atheas, who was conquered by Philip, king of Macedon.

15. The history of the Phrygians, who are said to be descended from Thogarme, the son of Gomar. Midas was one of their most ancient kings; he reigned soon after the deluge of Deucalion. After the death of Adrastus, who lived in the time of Crœsus, the royal house was extinct, and Phrygia became a province of Lydia.

16. The

16. The history of Phrygia Minor, or Troy. Dardanus and Teucer were its first kings, and Æneas its last.

17. The history of the Mysians. Olympus is situate in this country; and the first king of Mysia was also called Olympus in history. The last was Arius; though there are mention made of kings of Mysia in the time of the Attalian kings of Pergamus.

18. The history of the Lydians. Their first king was Mones, and their last Cræsus, who was vanquished by Cyrus.

19. The history of the Lycians. Their origin, and a great part of their history, belongs to fabulous times. One of their kings, named Cyberniscus, commanded in the fleet of Xerxes against the Greeks.

20. The history of the Cilicians. It is pretended that they drew their origin from Tarsis, the son of Javan, who peopled Cilicia, and gave his name to the city Tarsus. This people had kings at Thebes and Lyrnessus, who all bore the common name of Syennesis. Cilicia did not become a province to Macedonia till after the destruction of the kingdom of Persia.—And such were the principal ancient nations, of whom any history, though imperfect enough, has come down to us.

XII. The second part of ancient history contains, " The history of the other empires, monarchies, republics and lesser states, that have anciently

anciently subsisted in the world, and of whom no knowledge is to be had, but from profane writers." And among whom we consequently find more obscurity, less order, less connection, and less certainty. But, before we proceed to the analysis of these histories, let us here make some general reflections, that perhaps may not be without their use. If we consider the vast extent of the known part of the earth, and remember that it has always been divided into great, middling, and small states; and if we reflect on the immense number of mankind that must have there existed, and that the human race have constantly been divided into nations, governments, and colonies, more or less numerous, we must be surprized to find, in the general system of ancient history, which comprehends a space of 4000 years, so small a number of particular histories. It is therefore necessary to observe, that, in the first ages of the four quarters of the world, Asia alone was civilized; and, consequently, the first order of men was to be found in that country only. Europe and Africa were scarce discovered, or at most their borders, and the people who inhabited them, only were known. The center of Europe was as unknown as the center of Africa is at this day. That center is about the country which is now called Franconia; for, if we place one point of a compass on that part where stands the city of Nurenberg, and describe a circle with the other, we shall comprehend very nearly all our part of the globe.

globe. The septentrional regions were entirely unknown, though they were very populous. But all these inhabitants of Europe and Africa, especially those who lived toward the two poles, were nothing better than a sort of savages, without manners and without knowledge, ignorant of the use of letters, and, in a word, such as mankind in general are, without arts and sciences. The Romans discovered them by degrees, subdued them, and sent among them a sort of polishers, to make them more tame and tractable, and to inspire them with notions of humanity, as in our days we send missionaries into the southern countries as we discover them. The Romans bestowed on all these people the title of barbarians, which they right well deserved: they also sometimes sent their criminals amongst them, by way of banishment. Now, if we even knew the history of these people, it would not certainly be worth the while to write it or study it. For a history that affords no instruction becomes an object of mere idle curiosity, and is only an useless burden to the memory; it would perhaps be altogether as interesting to know the history of a colony of baboons, as such figures of men as these. On the contrary, it is of consequence to us to know the history of those polished nations who inhabited ancient Asia and its neighbouring countries, and, in general, of all civilized people; and of these we have sufficient accounts in the annals that are come down to us.

XIII. Ame-

XIII. America remaining undiscovered till the beginning of the fifteenth century, the knowledge of its first inhabitants cannot make any part of ancient history. For the rest, we must here observe again, that as the second part of ancient history, which is called profane, includes so many obscurities and fables, which preceded the real facts, Varro has divided time into three parts. The first comprehends obscure and uncertain time, which is, from the origin of the human race to the deluge of Ogyges, about the year of the world 2208; 1796 years before the common era, and 1020 before the first Olympiad. The second includes the fabulous time, and begins with the deluge of Ogyges, and continues to the Olympiads, that is, to the year of the world 3228, and 776 before the common era: this continued 1020 years. The third comprehends the historic time, and begins with the Olympiads, that is, in the year of the world 3228, and 776 before the vulgar era. It is called Historic, because, since the Olympiads, the truth of facts that have occurred has been confirmed by history.

XIV. The poets have also divided history after their manner, that is to say, by fictions. They distinguish, first, the golden age, which they attribute to Saturn and Rhea; the second is the silver age, ascribed to the reign of Jupiter. This age they extend to the time that tyrants appeared among the human race; who, to

render

render themselves powerful, oppressed mankind by violence and injustice. The silver age, therefore, must terminate with the time that Nimrod, the grandson of Cham, rendered himself terrible, built Babylon, and laid the foundation of the empire of the Chaldeans, about the year of the world 1771, and 115 years after the deluge. The third was the brazen age, which was, when rapacious men, possessed with the lust of dominion, endeavoured to reduce their brethren to a state of slavery. The siege and burning of Troy by the Greeks happened in this age, with which likewise the poets finish the time when those heroes they called demi-gods appeared upon the earth. The fourth age is that of iron, which began with the first Olympiad, that is, in the year of the world 3228. About this time Hesiod complains of living in an iron age; and Ovid, in the description he gives of it, says, that all sorts of crimes began then to prevail. They pretend it still continues; but we may say with the worldling,

Oh! le bon tems, que ce siecle de fer!

XV. As we comprehend, in the idea of ancient history, a continued series of all facts and events that have happened among civilized nations, from the creation of the world to the birth of Christ, being a space of about 4000 years, we are here to consider, under profane history,

1. The

(1.) That of the empire of *China*. They talk much of the chronology of this people, which according to Father le Compte*, includes more than 40,000 years from the foundation of their empire; but unless it can be clearly proved, that the Chinese have known the use of letters for 40,000 years past, we must regard their chronology as fabulous, chimerical, and altogether ridiculous: for there is no tradition, no other monument or voucher that can last so long. And and supposing the Chinese to have existed for so great a period, must there not have been other people upon the earth? Were not India, and all other countries adjacent to China inhabited? And must not these people have learnt from the Chinese, in 40,000 years, the use of letters? Is it possible, that the communication between neighbouring nations could be so far interrupted? The ridiculous fables likewise, with which the antient Chinese history is crowded, from beginning to end, confirm, in every rational mind, a contempt for their boasted chronology. Their most sagacious historians, moreover, commonly suppose that Fohi, their first king, mounted the throne 2252 years before the birth of Christ. The character they draw of this Fohi, is not unlike that of Noah, who may also very well be that Saturn of whom the poets talk, and who lived about the same time. Confucius the philosopher, a priest and legislator of the Chinese, flourished

Vol. I. p. 205.

about

about 550 years before the common era. In the year of Chrift 1279, the Tartars made themfelves mafters of this empire, and their family bore the name of Iven.

XVI. (2.) The hiftory of *Egypt*. The chronology of the Egyptians is altogether as extravagant as that of the Chinefe, and has no better foundation. The Chaldeans or Babylonians affigned myriads of years to their monarchy. The Egyptians, piqued at their pretenfions, would not yield them the preference in point of antiquity. Their priefts, and thofe they called fages, afferted that gods and demi-gods reigned in Egypt 42,984 years before their kings. It would be fome fatiffaction to know by what channel, or rather by what miracle, the knowledge of this has come down to our days, fuppofing it to be true. They have found means however to gain credit for thefe reveries with Diodorus Siculus, Herodotus, Manethon, and many others equally weak, credulous, and fond of marvellous relations. The indefatigable labours of that learned writer John Marfham, united with thofe of Ufher, and fome other able chronologers, have helped to diffipate, in fome degree, this real Egyptian darknefs, and to reduce the hiftory of this country, quite fabulous as it is in its origin, to a fyftem tolerably rational. This hiftory then is divided into dynafties, or races of fovereigns that have reigned in Egypt. Seven of thefe dynafties comprehend the reign of gods, from Vulcan to Typhon : nine,

the

the reigns of the demi-gods from Orus to the demi-god Jupiter. It is easy to conceive what credit is to be given to such history. Then come the obscure dynasties of the kings of Thebes, Thin, Memphis, and Heliopolis; and all this brings their history down to the time of Sesostris, or Sethosis, or Sesac, who reigned in the year of the world 3033. He made many conquests in Asia, and took Jerusalem in the fifth year of Rehoboam king of Juda. It is here that many historians quit Marsham, and follow the system of Usher. They begin the history of Egypt with the year of the world 1760; and consider this kingdom, 1st, as under unknown kings during 160 years; 2d. under six pastoral kings during 260 years, that is to the year 2180, when Amasis drove out these royal shepherds; 3. under 48 kings that are named Pharohs, during 1299 years, that is to the year 3479, when Cambyses king of Persia conquered Egypt; 4. under two Persian kings during 164 years to the year 3673, when Alexander joined Egypt to his other conquests; 5. under the Greeks, that is, under Alexander six years; 6. under 13 Ptolemies and Cleopotra the last queen of Egypt, during 294 years, which comes to the year 3974, when Augustus, after the death of Cleopatra, reduced Egypt to a province of the Roman empire, and lastly, 7. under the dominion of the Caliphs and Ottomans, from the time that Omar the second caliph, or heir of Mahomet, conquered Egypt in the year 637.

XVII.

XVII. (3) The history of the *Assyrian Monarchy*. We have already mentioned this history in the tenth section, but we cannot avoid speaking of it here, as one of the *four grand monarchies*, so called by way of excellence, and to which it is frequently the custom to reduce almost all ancient history. From this point of view, therefore, we regard the Assyrians, not as a particular nation, but as the sovereigns of Syria, Mesopotamia, Babylon, Persia, and, in a word, of all Asia except India. It is all these countries united that are comprehended under the name of Assyria the Great, which formed the empire of Ninus and Semiramis, and which is called for that reason *the first monarchy*; and sometimes it is also called the monarchy of the Babylonians, Chaldeans, Assyrians, &c. by which is always meant the same empire; and which they make to commence, for that reason, not before the year of the world 2737, with Ninus the son of Bel, and with Semiramis his consort; and to finish with Balthazar, who was slain by the soldiers at the taking of Babylon by Cyrus: and thus the Assyrian monarchy was overthrown, and passed to the Medes and Persians.

XVIII (4) The history of the *Persian Monarchy*, which is the second of those that were called Grand. This vast empire comprehended not only Persia, properly so called, and of which we have made mention in the second section, but almost all Asia, and sometimes also all the

the circumjacent country; seeing that Xerxes, after subduing all Egypt, came into Greece and took Athens. The historians make this grand monarchy to begin with Cyrus in the year of the world 3468, and to last 206 years under twelve kings, of whom Darius was the last, who being conquered by Alexander near Arbella, his estates and provinces passed to the dominion of the conqueror, and contributed to form the third grand monarchy.

XIX (5) The history of the *Grecian Monarchy*, which was the third grand monarchy. The history of this empire will be very difficult to explain, and still more to comprehend without making the following reflections. The Grecian monarchy did not properly subsist more than six years and ten months under the reign of Alexander surnamed the Great, who had already reigned six years over Macedonia, when he began to make himself master of the east; but, to conceive a clear idea of the manner in which so vast a monarchy was formed, it is necessary to begin by fully understanding the general history of Greece; then to study that of the kingdom of Macedonia, and after that to consider the life and conquests of Alexander in particular; to know the people, kingdoms, empires, and other states that he subdued, in order to form a monarchy almost universal; and lastly to know, how this immense monarchy was dismembered by his successors. The first object of inquiry therefore

is

is the history of the Greeks, the most curious and most important of all antiquity.

XX. Greece was so named from an obscure king called Græcus. Another king, who was named Hellen, gave to the Greeks the name of Hellenists. The different augmentations of this people have occasioned the learned to distinguish their history into four different ages, marked by the like number of important epochs. The first age comprehends almost 700 years, from the foundation of the small kingdoms of Greece to the siege of Troy. To this period belongs the foundation of Athens, Lacedemon, Thebes, Argos, Corinth and Sicyon; the atrocious act of the Danaides, the labours of Hercules, and, in general, all the exploits of the first heroes of Greece. The second age includes 800 years, from the Trojan war to the battle of Marathon. The third age continued only 158 years: it begins with the battle of Marathon, and ends with the death of Alexander. So many accomplished philosophers, orators and generals, never existed upon the earth at the same time, as during this period. The fourth age was not longer than the third; for, after the death of Alexander, the Greeks began to decline, and at last became subject to the dominion of the Romans.

XXI. Here we are to distinguish that which is called Great Greece, which comprehends some adjacent countries also with Greece properly so called.

called. It is very certain that never any country so small contained so many kingdoms and republics. They make the number amount to 49, among which are some whose names are scarce known. They are as follow: 1. Sicionia or Ægialia, 2. Leleg, 3. Messina, 4. Athens, 5. Crete, 6. Argos, 7. Lacedæmon or Sparta, 8. Pelasgia, 9. Thessalia, 10. Attica, 11. Phocis, 12. Locris, 13. Ozela, 14. Corinth, 15. Eleusina, 16 Elis, 17. Pilus, 18. Arcadia, 19. Egina, 20. Ithaca, 21. Cephalone, 22. Phthia, 23. Phocidia, 24. Ephyra, 25. Æolia, 26. Thebes, 27. Calista, 28. Ætolia, 29. Dolopa, 30. Oechalia, 31. Mycenæ, 32. Eubœa, 33. Mynia, 34. Doris, 35. Phera, 36. Iola, 37. Trachina, 38. Thresprocia, 39. Myrmidonia, 40. Salamine, 41. Scyros, 42. Hyperia or Melite, 43. The Vulcanian Isles, 44. Megara, 45. Epirus, 46. Achaia, 47. Ionia, 48. The Isles of the Ægean Sea, and 49. Macedonia.

XXII. All these states in fact flourished in Greece, and their united force resisted for a long time the attacks of their common enemies, especially those of the Persians, who were often roughly treated by the Greeks. We must not however form too grand an idea of all these kingdoms and republics. They were for the most part towns only surrounded by a small territory. The strength of the Greeks consisted more in their courage and conduct, than in the extent of their country, which however was extremely

tremely populous. They who would apply to the study of the Grecian history, should make it their principal endeavour to learn the different destinies of Scionia, Argos, Arcadia, Sparta, Athens, Corinth, Thebes, Mycenæ and Messene. For it is there that they will find great models of every kind, and they will there see that the politest genius, and the profoundest science may be united in one people, with the most noble and amiable valour.

XXIII. The kingdom of Macedonia made, as we have seen, part of Greece; its first king was Caranus, a native of Argina, and grandson of Hercules. This family reigned, in seventeen generations, till the time of Alexander. Philip, father of that celebrated hero, was an ambitious, able, warlike prince, and a great politician. He laid the foundation of that immense power which his son obtained by his numerous victories, and which will render him renowned to the end of time. The prophet Daniel compares him to a winged leopard, and in fact he flew from conquest to conquest; for in six years and ten months, he subdued Thrace, Greece, Egypt, a part of Arabia and Africa, Syria, Pamphylia, the two Phrygias, Caria, Lydia, Paphlagonia, Assyria, Susiana, Drangiane, Arachosia, Gedrania, Aria, Bactriana, Sogdiana, Parthia, Hyrcania, Armenia, Persia, Babylon, Mesopotamia, and India. All these extensive countries were added to Macedonia, and in the year of the world

world 3674; Alexander was declared king of Afia; when he made magnificent facrifices to his gods, and diftributed to his friends, his riches, cities and provinces, ftill however referving to himfelf the right of fovereignty. But he did not enjoy thefe great profperities long, for on May 22, 3681, he was taken off by a violent fever, in the thirty third year of his age.

XXIV. After the death of Alexander, thofe great men who had affifted in founding the Grecian or Macedonian monarchy, were the firft to overthrow and demolifh that coloffus. They divided the whole monarchy into ten provinces, whofe governors feemed to depend on four that were principal; thefe were Ptolemy who had Egypt; Seleucus, who reigned in Babylon and Syria; Caffander, to whom fell Macedonia and Greece; and Antigonus, whofe portion was Afia Minor. But this arrangement did not laft long, for each of them aimed at independence; and at length all the ftates, kingdoms, and provinces, that compofed the fucceffion of Alexander, and were governed by his fucceffors, paffed, one after the other, under the dominion of the Romans. All thefe ftreams, great and fmall, at laft fell into the ocean of the Roman monarchy, and were there loft.

XXV. (6.) The hiftory of the *Roman Monarchy*. The annals of mankind prefent nothing more grand than the Roman empire, as well

with

with regard to its power and extent, being incomparably more powerful and more extensive than any of the three former monarchies, as to the great men of every kind which it produced. To form a just and clear plan of this history, we must take matters from their origin, and transport ourselves to the country of the Latins. This country, the most celebrated of Italy, took its name from *latendo*, because Saturn, chased from his kingdom by his son Jupiter, came here to hide himself. The first inhabitants of this country were those called Aborigenes, the next were the Egueans, then the Ausonians, the Hernici, the Latins, the Rutuleans, and the Volcians. Latium is that country which is now called Campagna di Roma. Before the foundation of Rome, which became its capital, it was governed by kings, of whom are recorded,

1. Picus the son of Saturn, the first king of those Latins called Aborigines, because they were the original people of the country. He began his reign in the year of the world 2708, and reigned 37 years.

2. Faunus his son reigned 44 years.

3. Latinus his son reigned 34 years.

4. Æneas, the son of Venus, landed in Italy after the taking of Troy, and married Lavinia the daughter of king Latinus. He reigned after his wife's father only three years.

5. Ascanius, and fourteen other kings his successors, reigned in Latium till the time of Numitor

mitor and his brother Amulius, which was in the year of the world 3249.

6. Romulus and Remus, who laid the foundation of the city of Rome, and of a new empire.

XXVI. We may confider the Roman empire as under feveral different ftates.

1. Under feven kings from Romulus to Tarquin the Proud, during 245 years.

2. As a republic under the confuls during 465 years, that is to the year of the world 3960, when Cæfar began to make himfelf fovereign lord by the deftruction of liberty. Numberlefs actions of war and policy fignalized this period, and efpecially the three Punic wars, that is, thofe againft Carthage.

3. Under Julius Cæfar, who reigned with the title of perpetual dictator and imperator, or general of the army. He was affaffinated in the midft of the fenate. Auguftus and Pompey difputed the empire. Pompey fell. Auguftus reigned, and took the title of emperor. Jefus Chrift, the Saviour of the world, was born in Judæa; with this ever memorable period ancient hiftory ends. The hiftory of the middle age here begins, and comprehends the remaining part of the Roman monarchy, as we fhall fee in the following chapter.

XXVII. This is what we may, and ought naturally to comprehend under the idea of ancient

hiftory

history. To render this system however quite complete, it is proper to observe here, that, independent of the monarchies and empires which we have here enumerated, there have been in the world, during the first forty centuries, some other people and states, who though they have not arrived at that extreme power which constitutes empires of the first magnitude, and though they may not have produced events important enough to attract the attention of all future ages, yet are they notwithstanding worthy to be remembered, though it were only on account of the intimate connexion they have had with the four great monarchies; and consequently the study of their history becomes necessary. These people were,

In ASIA,

1. The *Idumeans* or Edomites, who inhabited the country of Seir, between Arabia, the gulph of Persia, and Judæa. The principal cities were Bozra and Petra. They united with the Jews in the time of Hircan, and had the same fate with them.

2. The *Arabians*, descended from Ismael. There is mention made of the kings of Arabia in the latter history of the Jews. In the reign of Trajan they became tributary to the Romans, but they regained their liberty, and at last submitted to Mahomet in the year 625; from which time their princes have been called caliphs. About the same time a party of these Arabs passed into Africa,

Africa, drove the Vandals from thence, and established themselves on the borders of Tunis.

3. The *Armenians*, whose country was anciently a province of Persia, and in that state fell under the government of the Macedonians. During the reign of Tigranes, the Armenians were conquered by the Romans, from which time they were governed by petty princes, and at last fell under the dominion of the Parthians.

4. The Amazons, who dwelt, as is supposed, in Cappadocia, and were originally Scythians. The first queens, of whom they talk, lived in the time of Adystus of Argos. The latter of them were conquered by Theseus, and the remains of this nation established themselves after that time beyond the river Tanais.

5. The *Carians*, who inhabited Asia Minor, and were anciently called Leleges. They were for some time subject to Minos, king of Crete; were afterward conquered by Cyrus, and at last submitted by degrees to the Ionians.

6. The *Odrises*, a people of Thrace.

7. The *Paphlagonians*, who dwelt between the Euxine sea and Galatia, and took their name from Paphlagon son of Phineas. They were conquered successively by Crœsus, Cyrus, and by the Romans, who, during the time of the emperor Dioclesian, incorporated that state with the province of Pontus.

8. The kingdom of *Pergamus*, whose last king, Attalus, left by his will his kingdom to the Romans.

9. The

9. The kingdom of *Bithynia*.

10. The kingdom of *Cappadocia*.

11. The kingdom of *Pontus*, which ended with Mithridates.

12. The kingdom of *Armenia*. These five small kingdoms were situate in Asia Minor.

13. The kingdom of the *Parthians*, whose kings were named Arsacidæ. It finished 126 years before the vulgar era, after Artaban IV. was killed by the Persians.

14. The kingdom of *India*, which took its name from the river Indus. Bacchus, they say, was the first who conquered it; the kings of Persia possessed a portion of it till the time of Alexander. Since his death the Indians have always had kings of their own.

In AFRICA.

XXVIII. (1) The *Carthaginians* were a colony of Phœnicians, who established themselves in the year of the world 3147, acquired a formidable power by their commerce, and possessed all the western coast of Africa. They were reduced by the three Punic wars to a Roman province.

2. The *Cyreneans* were a Grecian colony established in Africa.

3. The *Ethiopians*: who though they had always their own kings, yet their history is so connected with that of the Egyptians, as to make them inseparable.

4. The

4. The *Numidians*, who had always powerful kings. Masinissa and Jugurtha were formidable to the Romans, who neverthelefs reduced this kingdom at laſt to a Roman province.

In EUROPE.

XXIX. (1) The *Etruscans* in Italy, between the Tiber and Appenine mountains. This country was called Tuscia. They are said to have been originally Lydians. The Gauls, by their invaſions, obliged them to change their ſtation, and by degrees they became ſubject to the Romans.

2. The *Iberians* dwelt originally in Aſia. One of their colonies was eſtabliſhed on the coaſt of Spain, where they were oppoſed, firſt by the Carthaginians, and afterwards by the Romans.

3. The *Illyrians*, who inhabited the country that is now called Dalmatia, and ſome other parts. They had originally their own kings, but at length ſubmitted to the Roman yoke.

4. The *Britannic* iſles, or the kingdom of *Albion*. The firſt kings of theſe iſles were Britons. Julius Cæſar diſcovered, as we may ſay, theſe iſlands, and it was with much difficulty that the Romans maintained their dominion there.

5. The *Gauls*. Their country was divided into Ciſalpine and Tranſalpine. Cæſar reduced them to the Roman authority.

6. The *Pannonians*: who inhabited Hungary, Dalmatia, and European Turkey. They formed a powerful

a powerful nation, and were not reduced by the Roman emperors till very late, and did not remain any long time under their dominion.

7. The *Thracians*. A rough and warlike people, who inhabited the modern Romania; their firſt king was called Teres. This country was ſubdued by the ſucceſſors of Alexander. The Gauls overrun it ſoon after: but they were drove out by one Deuthes, whoſe ſucceſſors reigned tranquilly over this nation to the time of the emperor Veſpaſian.

XXX. Whoever ſhall apply to the ſtudy of ancient hiſtory according to the plan here laid down, we well hope, will be able to acquire a complete knowledge of it, eſpecially if they ſhall make a judicious choice of the beſt hiſtorians and moſt faithful annals that are ſtill remaining, of theſe remote, and very frequently obſcure ages.

CHAP. VI.

The HISTORY of the MIDDLE AGE.

I. AS we do not find, in the writers of universal hiftory, the limits of that period, which is comprehended under the term of Middle Age, either diftinctly or uniformly marked, we may be allowed to fix its bounds here, by two of the grandeft epochs in all hiftory, fuch as ftrike the mind and make the ftrongeft impreffion on the memory, and form at the fame time fo natural a divifion in hiftory, that the chronological order of facts becomes thereby more clearly and eafily conceived. We include therefore, in the middle age, thofe eight centuries which paffed between the birth of Chrift, and the re-eftablifhment of the Weftern empire by Charlemagne; who was crowned emperor at Rome on Chriftmas day in the year 800, by pope Leo III.

II. At the birth of our Saviour, Auguftus, the firft emperor that was acknowledged in that quality, and as fovereign, reigned over the Roman monarchy, the whole earth being under his dominion, except China and thofe countries that were either unknown, or too diftant to be included,

or

or inhabited by savage nations, or too inconsiderable to attract regard. All that was worth the trouble of conquering, and all whose history is worth the trouble of studying, was conquered, and in subjection to the Roman empire. The history of all the nations of the earth, during the middle age, is therefore included in the annals of the Roman monarchy: and when a people that was unknown, as for example, the Vandals, the Herulians, the Saracens, and others, appeared upon the theatre of the world, and made invasions or conquests in the dominions of the empire; it is the business of general history to explain the particular history of such people, as far as it is capable of explanation. For we cannot avoid confessing, that there reigns great obscurity in the middle age, and that there are many chasms in the histories of particular nations, who were either in subjection to the Roman empire, or at war with it.

III. The first objects, that offer themselves in the history of the middle age, are the Roman monarchy under forty-seven emperors, from Augustus to Theodosius the Great, who reigned over the known world for 395 years; and the translation of the seat of that immense empire from Rome to Constantinople. We then see the partition of that empire between the two sons of Theodosius, Arcadius and Honorius, and the establishment of the two empires, the Eastern and the Western, which arose from that division.

We

We learn, in the third place, the revolutions and the events that occurred in that part of the world which belonged to the empire of the East, of which Constantinople was the seat, and Arcadius the first emperor: and in the fourth place, we see all the revolutions and events that occurred in the dominions that made part of the Western empire, of which Rome was the capital, and Honorius the first emperor. This series of events continues, as we have said, till the time that Charlemagne re-established that empire, or rather when he formed a new one out of the ruins of the old. It will be necessary to give our readers a more circumstantial account of these matters, in order to enable them to form a clear idea of the knowledge they should endeavour to acquire of the history of the middle age.

IV. We have therefore to consider, in this age of 800 years, first, the Roman empire, under the following fortyseven emperors:

	Years.	Months.	Days.
1. Augustus, who reigned after the birth of Christ	15	0	0
2. Tiberius, his adopted son, who reigned	22	7	7
3. Caligula, son of Germanicus, reigned	3	9	28
4. Claudius, the son of Drusus,	13	8	20
5. Nero, his adopted son	13	8	0
6. Galba, the son of Servius Galba	0	6	7
7. Otho, the son of Salvius Otho	0	3	0
8. Vitellius, of an obscure family	0	8	2
9. Vespasian, the son of Titus Flavius Sabinus	9	6	2
10. Titus, the son of Vespasian	2	2	10

11. Do-

		Y.	M.	D.
11.	Domitian reigned	15	6	5
12.	Nerva, an old man, reigned only	1	4	9
13.	Trajan, a Spaniard	19	6	16
14.	Adrian reigned	20	10	29
15.	Antoninus	22	7	27
16.	Marcus Aurelius	19	0	0
17.	Commodus, the son of Marcus Aurelius	12	9	0
18.	Pertinax, the son of a brickmaker	0	3	0
19.	Didius Julianus purchased the empire, and reigned but	0	0	26
20.	Severus, who died at York, after reigning	17	8	3
21.	Caracalla and Geta succeeded their father Severus. Caracalla murdered his brother Geta at the end of one year and 22 days, and reigned, in all	6	2	5
22.	Macrinus reigned	1	1	26
23.	Heliogabalus	3	9	4
24.	Alexander Severus	13	9	0
25.	Maximinus, of Thrace	2	7	0
26.	Pupienus and Balbinus reigned scarce	1	0	0
27.	Gordianus	6	2	0
28.	Philip, with his son Philip II.	5	0	0
29.	Decius, surnamed Trajanus	2	0	0
30.	Gallus, with his son Volusianus	2	0	0
31.	Æmilianus	0	3	0
32.	Vallerianus and Gallienus	7	0	0
33.	Gallienus reigned alone, after his father, It was during this reign that the thirty tyrants arose.	8	0	0
34.	Claudius II, called the Goth, reigned	1	10	12
35.	Aurelian reigned	5	11	9
36.	Tacitus	0	6	20
37.	Probus, the son of a gardener	6	4	0
38.	Carus, with his two sons, Numerianus and Carinus, reigned altogether	2	0	0
39.	Dioclesian	18	0	0
40.	Constantius Chlorus,	2	3	0
41.	Constantine, surnamed the Great, reigned	30	9	27

He

Y. M. D.

He transferred the seat of the empire to Bizantium, and called it, after his own name, Constantinople. He also divided his empire into two parts, the East and the West. The East comprehended Hungaria, Transilvania, Valachia, Moldavia, Thrace, Macedonia, Pontus, Asia, and Egypt. The West contained Germany, Dalmatia, Sclavonia, Italy, Gaul, England, Spain, and Africa.

42. Constans, Constantius, and Constantinus, divided among them the empire of Constantine their father. This was a time of perpetual troubles and commotions, which lasted about   24   0   0
43. Julian, surnamed the Apostate, reigned but   1   8   0
44. Jovian, of Pannonia, reigned only   -   0   7   22
45. Valentianus reigned   -   -   11   8   22
46. Gratian, his son, divided the empire with Valentianus II. Gratian reigned   -   -   16   0   6
And Valentianus reigned 16 y. 5 m. 24 d.
47. Theodosius the Great reigned   -   -   16   0   20

V. This first period of the history of the middle age, under forty-seven Roman emperors, includes therefore 395 years, and comprehends, as we have said, the history of all nations, as all known parts of the earth formed Roman provinces, or were at war with that people; for their lust of dominion led them to attempt the conquest of every country they knew. Theodosius divided the empire between his two sons. Arcadius had that of the East, and continued his residence at Constantinople, as did his successors. This empire of the East lasted 1058 years, under seventysix emperors, to the time of Constantine Palæologus, who perished at the taking of Constantinople

ſtantinople by Mahomet II. in the year 1453; after the death of whom, this formidable empire paſſed under the dominion of the Ottomans. This firſt period of the hiſtory of the Eaſtern empire deſcends therefore from Arcadius to Nicephorus Logothata, the 29th emperor, who was elected by the army after the death of Irene, in the year 802 of the Chriſtian era, and to that period, this hiſtory belongs to the middle age. The ſecond period begins with that emperor, and ends with the taking of Conſtantinople. It comprehends the ſucceſſive reigns of fortyeight emperors, to Conſtantine Palæologus, during 641 years. This laſt period makes, properly, part of modern hiſtory, and may be very well ranged under that diviſion. But that we may not interrupt the regular ſeries, by being obliged to recur to it in the ſucceeding chapter, which will be otherwiſe ſufficiently long, we ſhall here bring it to a concluſion.

VI. The firſt period of the hiſtory of the Eaſtern empire, which belongs to the middle age, comprehends therefore the following reigns;

		Y.	M.	D.
1. Arcadius, who reigned	-	13	3	15
2. Theodoſius II.	-	42	2	28
3. Marcian	-	6	6	0
4. { Leo	-	17	0	0
{ Leo II. called the younger, reigned only		1	0	0
5. Zeno, of Iſauria,	-	17	0	0
6. Anaſtaſius, of Dyrrachium or Dicorus	-	27	3	3
7. Juſtin	-	9	0	23

8. Juſ-

ANCIENT HISTORY.   129

	Y.	M.	D.
8: Juſtinian, by whoſe order was made the Roman code, and to whom the famous Beliſarius was general, reigned	38	7	13
9. Juſtin II. called Curopalatis	10	10	20
10. Tiberius Conſtantine	6	10	8
11. Maurice of Cappadocia	19	3	11
12. Phocas	8	4	9
13. Heraclius	30	10	0
14. Conſtantine, his ſon	0	4	0
15. Heracleonas, the ſecond ſon of Heraclius	0	5	0
16. Conſtans II. the ſon of Conſtantine	17	0	0
17. Conſtantine, called Pagonatus, or Longbeard	17	0	0
18. Juſtinian II. his ſon	10	0	0
19. Leona, empreſs, reigned	3	0	0
20. Abſimarus Tiberius, whoſe reign was one ſcene of troubles	13	0	0
21. Philippicus Bardanes	2	9	7
22. Anaſtaſius	1	3	0
23. Theodoſius III. ſcarce	1	0	0
24. Leo of Iſauria, called Ichonomachus	24	2	25
25. Conſtantine V. ſurnamed Copronymus	34	2	26
26. Leo IV. his ſon	5	0	0
27. Conſtantine VI. called Porphyrogenitus, and 28. Irena, his mother, reigned	18	0	0

And here finiſhes the firſt period of the hiſtory of the Eaſtern empire, and the middle age, with the year of our era 801.

VII. The ſecond period of this empire (which makes part of modern hiſtory) contains the following reigns:

	Y.	M.	D.
29. Nicephorus Logothata, who reigned	8	0	0
30. Michæl Curopalatis	2	0	0

31. Leo

		Y.	M.	D.
31.	Leo of Armenia	7	5	0
32.	Michel II. called the Stammerer	8	9	0
33.	Theophilus, his son	12	3	20
34.	Michel III. son of Theophilus	11	1	9
35.	Basilius of Macedonia	18	10	7
36.	Leo VII. called the Philosopher	25	0	0
37.	Alexander, his brother, about	1	0	0
38.	Constantine VII. Porphyrogenitus, reigned near	47	0	0
39.	Romanus Porphyrogenitus, his son	2	15	0
40.	Nicephorus Phocas	6	6	0
41.	John Zimisces	5	11	0
42.	Basilius and Constantine, brothers, died one after the other, having reigned more than	50	0	0
43.	Romanus reigned	5	6	0
44.	Michel of Paphlagonia	6	6	0
45.	Michel Calaphatus reigned only	1	0	0
	Zoë, the widow of Michel of Paphlagonia assumed the government, and reigned, with her sister Theodora	1	0	0
	And then married.			
46.	Constantine Monomachus, who reigned with Zoë and Theodora	12	0	0
47.	Theodora reigned after them, alone	1	9	0
48.	Michel the Warrior, an old man, reigned	1	0	0
49.	Isaac Comnenus	2	3	0
50.	Constantine Ducas	7	6	0
51.	Eudoxia, his widow, reigned with John, brother of Constantine, and the three sons which he had by his wife, Michel, Andronicus, and Constantine, about	1	0	0
52.	Romanus Diogenes married Eudoxia, and reigned	3	8	12
53.	Michel Ducas	6	6	25
54.	Nicephorus II.	3	6	0
55.	Alexis Comnenus	37	4	15

In his time, that is, at the beginning of the twelfth century, commenced the famous croi-

sades

Ancient History. 131

	Y.	M.	D.

sades of the Occidental Christians against the Turks and Saracens, in the East.

	Y.	M.	D.
56. John Comnenus	24	8	0
57. Manuel Comnenus	37	5	0
58. Alexis II. his son	3	0	0
59. Andronicus Comnenus	2	0	0
60. Isaac, of the house of Angelus Comnenus	10	0	0
61. Alexis III. his brother	8	3	0
62. Alexis Mirtillus reigned only	0	2	15

At this time there began to be two seats of empire, one at Adrianopolis, by Theodorus Lascaris, and the other at Trebizond, by Alexis Comnenus.

63. Baldwin, a Frenchman, made himself master of Constantinople, and was crowned emperor, in the year 1204. But he reigned only   0 11 0

INTERREGNUM.

	Y.	M.	D.
64. Henry, count of Flanders, brother of Baldwin, succeeded him, and reigned	10	0	0
65. Peter de Courtenai, count of Auxerre, reigned	5	4	0
66. Robert, his son	7	0	0
67. Baldwin II. the son of Robert, after reigning was drove out of Constantinople by Michel Palæologus, the tutor of John and Theodore Ducas III. the sons of Theodore Ducas II. who had reigned at Adrianopolis.	30	0	0
68. Michel Palæologus made himself emperor, and reigned	22	0	0
69. Andronicus II. his son	42	0	0
70. Andronicus III. Palæologus	13	0	0

71. John V. Palæologus reigned under the tutelage of

72. John VI. Cantacuzenus, who usurped the whole authority, but at last gave his daughter in marriage to John, and, after having reigned — — — 14 6 15

with

                                                    Y. M. D.
with his son-in-law and pupil, was obliged
to abdicate, and turned monk.
John V. reigned alone   -   - 28 0   0
73. Andronicus IV. Palæologus reigned -   3 0   0
74. Emanuel Palæologus   -   -   - 31 0   0
75. John VII. Palæologus   -   - 27 0   ●
76. Constantine XIII. or, according to others,
    XV. and last emperor of Constantinople, was
    pressed to death amidst the multitude, at the
    taking of that city by Mahomet II. in the
    year 1453, after having reigned about -   8 0   ●

VIII. It were to be wished that we could cover with a thick veil the whole history of the Eastern empire, and conceal from the eyes of youth those horrors with which it is crowded from beginning to end. All these emperors, unworthy of so august a title, were either egregious dolts, or execrable villains; who acquired the diadem, and maintained it, by the blackest treasons and murders. A stream of blood flowed incessantly. Continual instances of poisoning, putting out of eyes, and other like horrors. No traces of genius or of virtue. This part of history ought to be made known merely to inspire a just aversion to guilt: what is still more deplorable, all these crimes were committed under the shadow of religion, or rather fanaticism and superstition. We shall see in the history of the church, by what unlucky schism Christianity was, so to say, torn asunder, and divided into the Greek and Latin churches. Constantinople adopted
                                                    the

the dogmas and rites of the Greek church, and Rome, the dogmas and rites of the Latin.

IX. They who would make a thorough study of the history of the middle age, should there include the particular histories of such people as were in subjection to the Eastern empire, or against whom its emperors waged war. The bounds of this work will not permit us to enter into so large a detail; but when, in the next chapter on modern history, we come to treat of the Ottoman empire, we shall not forget to inform our readers who those Turks were that took Constantinople under Mahomet II. and made it the seat of their empire. It only remains here to say a few words on the kingdom of Jerusalem, the emperors of Trebizond, and those of Adrianopolis.

X. The kingdom of Jerusalem continued only 88 years, under nine kings; that is, from the year 1099, when the Christian army took Jerusalem from the sultan of Egypt, to 1187, when Saladin, sultan of Syria and Egypt, retook it from the Christians. There reigned, during that time,

	Y.	M.	D.
1. Godfrey of Bouillon, scarce	1	0	0
2. Baldwin succeeded his brother, and reigned	18	0	0
3. Baldwin II. succeeded his cousin, and reigned	12	0	0
4. Foule, count of Anjou, married Beatrix, the daughter of Baldwin II. and reigned	11	0	0
5. Baldwin III. who succeeded his father	21	0	0

6. Amau-

	Y.	M.	D.
6. Amaurus, count of Afcalon, fucceeded his father	10	0	0
7. Baldwin IV. the Leper, fucceeded his father	13	0	0
8. Baldwin V. reigned only under the protection of	0	3	0

9. Guy of Lufignan, when the city of Jerufalem was taken by affault, 2 Oct. 1187, and all the Holy Land paffed into the hands of the infidels.

XI. Colchis, or the province of Trebizond, was in the poffeffion of Alexis Comnenus, with the title of principality, under the emperors of Conftantinople, when the French took that capital, in the year 1204. Alexis, feeing Conftantinople in the hands of the French, made himfelf fovereign of Colchis, without however affuming the title of emperor; nor did the two princes who fucceeded him. It was the fourth fovereign, John Comnenus, who ufurped that title. Trebezond therefore had nine emperors, who were, 1. John Comnenus; 2. Alexis Comnenus; 3. Bafil Comnenus; 4. Bafil Comnenus II. 5. N. Comnenus, the natural fon of Bafil II. 6. Alexis Comnenus II. 7. Alexander Comnenus; 8. John Comnenus II. and 9. David Comnenus, who furrendered all Colchis to Mahomet II. a conqueror too powerful for him to withftand. Thus ended the empire of Trebizond, after having lafted 257 years.

XII. The fmall empire of Adrianopolis was founded in 1204, by Theodore Lafcaris, who had

had married Ann Comnenus, the daughter of Alexis Comnenus. It continued only 60 years, under four emperors, who were,

	Y.	M.	D.
1. Theodore Lascaris, who reigned	18	0	0
He married his daughter Irene to			
2. John Ducas, who succeeded him, and reigned	33	0	0
3. Theodore II. succeeded his father, and reigned	4	0	0
4. John, his son, succeeded, and reigned one year and some months with his brother Theodore. Michel Palæologus, their tutor, caused them to be murdered, and then joined the empire of Adrianopolis to that of Constantinople, of which he had already made himself master. And thus we have given a slight sketch of the most important events that occurred in the Eastern empire.			

XIII. The fourth part of the history of the middle age comprehends the history of the Western empire, from the partition of the Roman monarchy, that is, from the year 395, to the year 800, of the Christian era; and in which we distinguish the following epochs:

1. The emperors of the West, to the year 475.

	Y.	M.	D.
1. Honorius, who died at Ravenna, after reigning	28	0	0
2. Valentinian III. the son of Constantius, reigned	30	0	0
Attila, king of the Huns, ravaged Italy: the empire of the West declined, and the imperial seat was almost always, after this, at Ravenna.			
3. Maximus usurped the empire; but he was soon cut in pieces by the Romans, and thrown into the Tiber; he reigned only	1	0	•

Genseric,

Genseric, king of the Vandals, whom Eudoxia, widow of Valentinian, had called from Africa, entered Rome, and pillaged that city for 15 days. The Western empire is destroyed. Africa is possessed by the Vandals; Spain by the Visigoths; Gaul by the Franks; the British Isles by the Picts, English, and Saxons; and Italy by the Lombards. The princes who succeeded were rather pretenders to empire than emperors. However, we must not omit them.

		Y.	M.	D.
4. Avitus reigned		1	2	8
5. Majorian		4	4	2
6. Severus		3	8	27
7. Anthemius		5	2	28
8. Anicius, called Olibrius, reigned		0	7	16
9. Glicerius abdicated, after he had reigned		1	3	21
10. Julius Nepos reigned		1	2	0
11. Romulus Augustulus, son of Orestes, was the last Roman emperor that was acknowledged at Rome; he was dispossessed by Ordoacer, king of the Heruls, after having reigned only		0	10	5

The West was, after this, without emperors for 324 years, that is, to the time of Charlemagne. Thus the ancient Roman empire, that was so formidable under the first Augustus, was reduced to a mere shadow of existence under Augustulus: this sovereignty, which began very lowly under the first Romulus, ended still more insignificantly under the last Romulus; and was lost like a rivulet that runs into the ocean.

XIV. In order to have a just conception of the history of the middle age in general, and of that of the empire of the West during its decline and

and diffolution, in particular, it is indifpenfably neceffary to acquire fome knowledge of thofe ferocious people, who, in the fourth and fifth centuries, over-ran all Europe, and penetrated even into Africa. But as all thefe people were barbarians, a kind of favages, without arts or fcience, even ignorant of the ufe of letters, and who had always been in a manner vagabonds upon the earth, without city or country, it is evident that they could have no annals, and that all we can fay of their origin and their hiftory muft be a mere collection of conjectures. It is, moreover, impoffible for us to enter here into the labyrinth of learned inquiries; we muft therefore content ourfelves with giving the names of thefe people, and merely informing our readers of what they ought to inquire after in this part of univerfal hiftory.

XV. The great and memorable *migration of people* happened toward the clofe of the fourth, and in the fifth centuries of the Chriftian era. A numerous fwarm of unknown and barbarous nations came, in part from the north, partly from the Palus Mæotis, and partly from the Eaft, by Hungary and Pannonia, and entered the provinces that formed the dominion of the empire. Thefe people gravitating on each other, to ufe the expreffion, conftantly impelled all that were before them, till they at laft penetrated the fouthern confines of Europe and Italy itfelf; where, meeting with a weak refiftance only, they

put

put an end to the succession of Roman emperors, and to their monarchy. The principal of those wandering and warlike people were,

1. The Visigoths, who appeared under the conduct of their king, Alaric.

2. The Ostrogoths, who became famous under their king Theodoric, who conquered Italy, and whose descendants possessed it for a long time.

3. The Vandals.

4. The Alains.

5. The Suevians.

6. The Heruleans, who were led by their king, Odoacer.

7. The Huns, of whom Attila was chief.

8. The Longobards, or Lombards.

9. The Picts.

10. The Scoti, or Scotch.

11. The Slavi, or Esclavonians.

12. The Gepideans and Avarians.

All these people flocked, one after the other, from the grand seminary of mankind, that is, from the most northern provinces of Europe, and even of Asia: as Norway, Sweden, Russia, and perhaps Siberia and Tartary also. The most part of the names they bore are analogous to the modern low Saxon, or seem to be derived from it. The Goths, for example, signify, in that language, Good people: the Quades, the Bad: the Huns, Dogs: the Slavi, Slaves: the Longobards, perhaps Longbeards; and so of the rest. It is apparent, that the greatest part of these

these people came from those countries that make part of Low Saxony.

XVI. All these people are frequently confounded with each other in history; and frequently, also, the same people was divided into different governments, which had each a particular name. All this has produced a chaos very difficult to be reduced into any order. The greatest satisfaction is, that it is of very little importance, to the present inhabitants of civilized Europe, to know the particular histories of all these barbarians; and that it is of no consequence if we do sometimes err in these matters. But it is not a matter so insignificant to know the history of those who have made a conspicuous figure in the world, who have either founded or possessed grand sovereignties in Europe; and especially those who succeeded the emperors of the West, and became possessed of the ruins of their monarchy. It is with this view that we shall here treat of the history of the empire of the West, from the death of Romulus Augustulus to the time of Charlemagne: and when, in the history of empires, kingdoms, and other modern states, we shall have occasion to speak of their origin and antiquities, we shall endeavour to investigate the kind of establishment that these wandering people instituted in each one of them in particular.

XVII. It

XVII. It remains therefore to consider here the state of the Western Roman empire under nine kings; one of the Heruleans, and eight of the Ostrogoths; during ninety-two years. According to common opinion, the Goths came from Scandinavia, a peninsula which is now inhabited by the Swedes and Norwegians. After having roved some time on the borders of the Baltic Sea, they passed into Scythia, and established themselves along the borders of the Euxine Sea. They who advanced the furthest towards the east were called Ostrogoths, that is, Eastern Goths; and they who dwelt toward the west were named Wisigoths, or Western Goths. In the year 476,

		Y.	M.	D.
1. Odoacer, king of the Heruleans, made himself master of Rome, drove away Augustulus, and called himself king of Italy. He reigned	-	16	6	0
2. Theodoric, king of the Ostrogoths, reigned	-	33	6	0
3. Athalaric reigned - -	-	8	0	0
4. Theodahat - -	-	2	0	0
5. Witiges - -	-	4	0	0
During this reign Belisarius, general of the emperor Justinian, had well nigh drove all the Goths out of Italy.				
6. Theobald reigned - -	-	1	0	0
7. Alaric - - -	-	0	3	0
8. Totila - - -	-	6	0	0
9. Tejas, the last king of the Goths in Italy. After having taken many cities during the absence of Belisarius, he made himself master of Rome, and pillaged it for forty days; but, after reigning about -	-	1	0	
He was vanquished by Narses, whom the em-				

peror

peror Juſtinian had ſent into Italy: and thus ended the kingdom of the Oſtrogoths. Italy remained in poſſeſſion of the emperor of the Eaſt, and Narſes obtained the government; but the Romans ſent great complaints to Conſtantinople, againſt that eunuch. Juſtinian recalled him in anger; but, inſtead of obeying, he ſent ſecretly to Albion, king of the Lombards, who paſſed into Italy, and there founded a kingdom that laſted till the time of Charlomagne.

XVIII. The laſt ſtate of the Weſtern empire, in the middle age, comprehends therefore the reigns of twenty-three Lombard kings, during 205 years. The Lombards were alſo a northern people, who firſt eſtabliſhed themſelves on the ſouthern borders of the Baltic Sea, and advancing by degrees more toward the ſouth, at laſt penetrated Italy, and there founded their kingdom in Ciſalpine Gaul; which was after named Lombardy.

	Y.	M.	D.
1. Albion entered Italy, and there reigned	3	6	0
Juſtin, emperor of the Eaſt, ſent Longin to Ravenna, as his exarch.			
2. Celphis reigned	1	5	0
After his death there was an interregnum, which laſted	10	0	0
3. Antarit reigned	5	6	0
4. Agilulf	26	0	0
5. Adelwald reigned alone, after his father	9	0	0
6. Ariowald reigued	12	0	0
7. Rotharis	16	4	0
8. Rodoald	6	0	0

9. Ari-

		Y.	M.	D.
9. Aribert	- - - -	2	0	0
10. Gondebert, and his brother				
11. Berthier, reigned, together, only	-	1	3	0
12. Grimoald - -	-	8	0	0
13. Garibaud, his son - -	-	0	3	0
Berthier returned to Pavia, where he was again acknowledged as king, and reigned	-	18	0	0
14. Cunibert reigned, after his father Berthier		12	0	0
15. Luitbert, his son, reigned only	-	0	8	0
16. Racombert, duke of Turin, reigned also	-	0	3	0
17. Aribert II. reigned -	-	8	0	0
18. Aufprand died at the end of	-	0	3	0
19. Luitprand, his son, reigned -	-	31	7	0
20. Hildebrand, his grandson, only	-	0	7	0
21. Rachis - -	-	5	6	0
22. Aftulf - -	-	6	0	0
23. Didier, duke of Etruria, reigned	-	17	0	0
He was conquered in the year 781 by Charlemagne, and the kingdom of the Lombards ended with him.				

XIX. After Narses had revolted, and while Albion was busied in founding the kingdom of the Lombards, the emperor of the East, Justin II. sent Longin to Ravenna, to endeavour to establish his affairs in Italy, and to promote his interest. But Longin made himself duke of Ravenna, and assumed the title of *Exarch*, by which is meant, *without superior*. They commonly reckon sixteen exarchs, who reigned at Ravenna during 184 years, to the time of Aftulf, the last king but one of the Lombards, who took that city, and put an end to the exarchy. These exarchs (whom some also name vicars, or prefects) were,

		Y.	M.	D.
1. Longin, who governed	-	15	0	0
2. Smaragdus	-	3	0	0
3. Romanus	-	11	0	0
4. Callenic,	-	4	0	0
Smaragdus, in his age, governed again	-	9	0	0
5. John Remiges	-	4	6	0
6. Eleuthera	-	3	0	0
7. Isaacius	-	23	0	0
8. Theodore Calliopas	-	8	0	0
9. Olympius	-	3	0	0
Theodore Calliopas is re-established, and governs	-	34	0	0
10. Theodosius, or Theodore, governed	-	0	6	0
11. John Platini, or Platon	-	15	0	0
12. Theophilactus	-	8	0	0
13. John Risocop, or Trisocop	-	5	0	0
14. Scholasticus	-	10	0	0
15. Paulus	-	2	0	0
16. Eutichius, the last exarch	-	24	0	0

Who, after he was drove from Ravenna, retired to Constantinople.

XX. In proportion as the Roman empire in the East, and more especially in the West, declined, there arose in the world, but principally in Europe, several kingdoms and other independant states, the revolutions of which belong to the history of the middle age. Thus, in the year 420, Pharamond reigned as king in France. In the year 408 the Goth and Vandal kings reigned in Spain, and so of the rest. But as these establishments include the origin of those

mo-

monarchies and states which now exist, it is more eligible to make the account of them precede the histories of modern nations, in order to prevent any interruption in the series of those histories, and all embarrassment in the study of them. The next chapter, therefore, will contain all that relates to this matter in the middle age.

XXI. But as the kings of the Vandals in Africa had, in the fifth and sixth centuries, much concern in the affairs of Italy, and as their empire disappeared before the end of the middle age, so that we shall have no further occasion to mention them, it appears necessary to give the chronology of the kings of that nation in this place. The kingdom of the Vandals in Africa lasted 108 years, under six kings, who were,

	Y.	M.	D.
1. Genseric, king of the Vandals, who founded that kingdom in the year 427, and reigned	48	0	0
2. Huneric, son-in-law of Valentinian III. succeeded his father, and reigned	8	0	0
3. Gondebaut, the grandson of Huneric, reigned	11	0	0
4. Trasimond succeeded her brother, and reigned	26	0	0
5. Hilderic reigned	8	0	0
6. Gilimer was put in the place of Hilderic, who had been unjustly deprived of the throne. He reigned	9	0	0

And was vanquished by Belisarius, in the year 534. And thus ended the kingdom of the Vandals.

Here we shall conclude our sketch of the history of the middle age: very happily, if we have in any degree difperfed that thick darkneſs with which it is furrounded; and have given a regular and clear plan, at leaſt, by which it may be fuccefsfully ſtudied.

CHAP. VII.

MODERN HISTORY.

CHARLEMAGNE appeared upon the earth. He was the ſon of Pepin, mayor of the palace of Childeric III. and the laſt king of France, of the Merovingian family. In the year 751, Pepin himſelf was made king, and died in 768. Charles was born in 741. He was of German extraction, of the nation called Francs, and was born in Germany. He eſtabliſhed the ſeat of the empire he founded at Aix-la-Chapelle: he and his courtiers ſpoke German; and the public acts were wrote in German. He ſubdued the other tribes of his nation, converted them to Chriſtianity, and taught them to read and write their native language. When he came to the throne of France, in the year 768, Conſtantine V. was emperor of the Eaſt, and, after him, Leo IV. Constantine V.I. the empress Irene, and Nicephorus. Italy was in the power of Astulf, king of the Lombards. Stephen IV. and, after him, Adrian I. and Leo III. were popes. Germany contained many nations that were but little civilized: Wittekind was chief of the Saxons. Winceslaus, and, after him, Crzezonisle, reigned in Bohemia. Gotric, or Sigefroi, was king of Denmark. Biorno III. and, after him, Alaric III. reigned in Sweden. Lescus I. was duke of Poland. The Saxons were masters of England, and had there established several small kingdoms, which were united in 801 under Egbert, first sole king of that country. Fergus, and, after him, Solvathius, Achaius, and Congallus, reigned in Scotland. Aurelius, and, after him, Silon, Mauregat, Veremond, and Alphonso the Chaste, reigned in the Austrias and the kingdom of Leon. The rest of Europe was entirely barbarous; and what they called civilized was not much better. Such was the state of Europe when Charlemagne, by the death of his father, became king of France. But this hero soon made new acquisitions, bearing in one hand the sword, and in the other the promises of the gospel. By the extinction of the kingdom of the Lombards in 773, he got possession of all Italy. By conquering the Saxons, and by converting them to Christianity, he became master of all Germany. By the election of the Roman people in 800, he obtained the empire of the West, with the title of Emperor; and a short time before his death, in 809, was very near adding to his vaſt dominions the kingdom of Spain.

II. It is therefore with the advancement of this monarch to the imperial dignity, with the re-eſtabliſhment of the empire of the Weſt, in the firſt year of the ninth century, that commences what is called Modern Hiſtory. The face of Europe was changed. It became at once Chriſtian and civilized. It was then that modern kingdoms, republics, and ſtates, were either founded, or acquired their true conſiſtence. This laſt age of the world, down to the preſent day, contains 965 years. The means by which the Divine Providence thought proper to civilize Europe, and almoſt all the other nations of the known earth, during that period; the ſucceſſive progreſs of the arts and ſciences; the uſeful inventions of every kind; the degree of perfection to which manufactures and commerce have been carried; the diſcovery of a new world; the eſtabliſhment of poſts and public banks, and of every kind of intercourſe between mankind; the improvement of navigation, and a thouſand like objects, require as much to be clearly inveſtigated and explained in modern hiſtory, as do the politics of kings, the ſtratagems of miniſters, the exploits of heroes, and the revolutions of kingdoms. It is therefore our buſineſs here to preſent our readers with a clew to this labyrinth, but we truſt they will not expect that we ſhould conduct them thro' all its minute windings and receſſes, by entering into a description of those small states that may be called miniatures of government. We shall therefore confine ourselves to an inquiry into the state of the following nations.

III. *I. The new Empire of the West, called the Holy Roman Empire.* 1. Before the interregnum: 2. During the interregnum: And, 3. After the interregnum.

Before the Interregnum.

1. Under nine Carolovinian emperors, or those who were the descendants of Charlemagne; to wit, 1. Charlemagne; 2. Lewis I. the Debonnaire; 3. Lothario; 4. Lewis II. 5. Charles the Bald; 6. Lewis III. the Stammerer; 7. Charles the Grofs; 8. Arnold; 9. Lewis IV. called the Child, who died without an heir in the year - - - - 912
2. Under six Saxon Emperors: that is, 1. Conrad I. Duke of Franconia. 2. Henry I. called the Fowler. 3. Otho I. called the Great. 4. Otho II. refused. 5. Otho III. called the Marvellous. 6. Henry II. surnamed the Saint, who died in the year — 1024
3. Under five Franconian Emperors, who were 1. Conrad II. the Salic. 2. Henry III. the Black. 3. Henry IV. 4. Henry V. 5. Lothario, who died in — 1137

4. Under

4. Under six Suabean Emperors, to wit, 1. Conrad III. of Suabia. 2. Frederic Barbaroſſa. 3. Henry VI. called the Severe. 4. Philip. 5. Otho IV. 6. Frederic II. who was poiſoned in the year — — 1250

During the Interregnum.

This was a time of trouble and confuſion that laſted twenty-three years; and during which, 1. Henry Raſpo of Thuringia. 2. Conrad IV. of Suabia. 3. William, Count of Holland. 4. Richard I. King of England. 5. Alphonſo X. of Spain: and, 6. Ottocar of Bohemia, were elected by different factions; or pretended to the empire, and endeavoured to attain it, either by cabals, or by force of arms; whilſt Conradin, Charles of Anjou, Mainfroi, and the Popes, excited a thouſand troubles in Italy. This interregnum at laſt ended in the year 1273

After the Interregnum.

1. Under twelve Emperors of divers houſes, choſe by the electors, 1. Rodolph, Count of Hapſbourg. 2. Adolph, Count of Naſſau. 3. Albert of Auſtria, called the One-eyed. 4. Henry VII. Count of Luxembourg. 5. Lewis V. of Bavaria. 6. Frederic III. of Auſtria, called the Handſome, who diſputed the empire with him; and, after the death of Lewis, Edward III. King of England; Frederic the Severe, Mar-

Margrave of Misnia; and Gunter, Count of Schwartzbourg, were elected emperors, without being able to get possession of that dignity, which fell at last on, 7. Charles IV. of Bohemia. 8. Winceslaus, King of Bohemia. 9. Frederic of Brunswic. 10. Robert of Bavaria. 11. Jocelin of Moravia: and, 12. Sigismond, son of the Emperor Charles IV. King of Hungary, who died in - - 1437

2. Under thirteen Emperors chose by electors from the house of Austria; to wit, 1. Albert II. 2. Frederic IV. 3. Maximilian I. 4. Charles V. 5. Ferdinand I. 6. Maximilian II. 7. Rodolph II. 8. Matthias. 9. Ferdinand II. 10. Ferdinand III. 11. Leopold. 12. Joseph: and, 13. Charles VI. who died in ——— ——— 1740

3. Under the Emperor Charles VII. Elector of Bavaria, who died in ——— 1744

4. Under the Emperor Francis I. Duke of Lorrain, and Grand Duke of Tuscany, who died in 1765

IV. *II. The empire of the East, possessed by the Sultans, or Turkish Emperors, or the Ottoman Porte.* This history divides itself naturally into two parts. In the first we are to investigate the origin of the Turks or Ottomans; and the fate of that people till the time of Mahomet II. who took Constantinople, and there fixed the seat of his dominion. In the second we are to bring the history of the Ottoman empire,

empire, from Mahomet II. down to the present time.

In the first part we shall see, that the Arabs or Saracens, who were a people descended from Ishmael the son of Abraham and of Hagar, inhabited the country which is called Arabia, from the word Araba, which signifies solitude. These Arabs are also sometimes called Ishmaelites, sometimes Agarenians, and sometimes Sarrazins, from the word Saraz, which signifies to steal; because this people traversed the country in order to rob on the highways. In 571 the false prophet Mahomet was born among them, and taught them a new religion, which they followed, as we shall see in the next chapter. Mahomet, who was at once a prophet, a legislator, and a conqueror, made himself sovereign of the Saracens or Arabs. The successors of Mahomet bore the title of Caliphs. About a hundred years after the death of Mahomet, a people of Scythia, named Turks, came by the Caspian Sea, past Mount Caucasus, and established themselves in that country, which is now called Georgia, Turcomania and Diarbeck. The Saracens at first waged war with these new comers; but about the middle of the eighth century they made peace, and incorporated with them, on condition that the Turks should embrace the Mahometan religion, and join with them in fighting against the Christians, who were come to molest them, even in Asia. The word Turk signifies a shepherd or peasant. It has effaced

faced that of Saracen and Arab. These two people therefore united, formed only one nation, and gave themselves the title of Muſſulmans, or true believers. The Caliphs, ſucceſſors of Mahomet, extended their dominions on every ſide. 1. Ottoman, whoſe origin and time of birth is uncertain, made himſelf maſter of the power and territories of all the other Caliphs and Soldans who then reigned in the Eaſt: he laid the foundation of that empire, which is called Ottoman, from his name, and took the title of Sultan. This happened in the year 1303. His ſucceſſors were, 2. Orchan: 3. Amurath: 4. Bajazet: 5. Izazebel: 6. Solyman: 7. Moſes: 8. Mahomet: 9. Amurath II. 10. and laſtly, Mahomet II.

V. In the ſecond part of the hiſtory of the Ottoman empire, we ſee Mahomet II. overthrow the empire of the Greeks in the Eaſt, make himſelf maſter of Conſtantinople, and there eſtabliſh the ſeat of his monarchy, and take to himſelf the title of Emperor and Grand Seignior. This great event happened on the 29th of May, in the year 1453. The ſucceſſors of Mahomet II. were, 1. Bajazet II. 2. Selim: 3. Solyman II. 4. Selim II. 5. Amurath III. 6. Mahomet III. 7. Achmet: 8. Muſtapha: 9. Oſman: 10 Amurath IV. 11. Ibrahim: 12. Mahomet IV. 13. Solyman III. 14. Achmet II. 15. Maſtapha II. 16. Achmet III. 17. Maſtapha III. which makes in all, from the firſt period to Mahomet II. ten
Sul-

Sultans: and in the second period, from Mahomet II. to our days, seventeen emperors or Grand Seigniors.

VI. After having thus brought down the history of the two grand empires of the East and West, to our own days, we may study to advantage the history of all other empires, kingdoms, republics, and modern states, by following the geographic order in which they present themselves to us in the map of Europe; beginning with the west, and advancing toward the east, till we come to Asia, Africa, and even to America, that we may learn the histories of the people who at this day inhabit those parts of the world. And in this manner we begin with learning,

VII. III. *The history of Portugal*; which is divided into the following epochs: 1. The origin of the Lusitanians; the description of ancient Lusitania, and of its inhabitants. 2. The first part of the history of the Lusitanians, to the year of Rome 607. 3. Their state and conduct under the Roman government, from the year of Rome 607 to the year of Jesus Christ 395. 4. The manner in which that country was invaded by the northern barbarians; and what passed to the year of Jesus Christ 800. 5. The fate of Lusitania during modern times, to the year 1075. 6. The government of the Moors in Portugal. 7. The erection of Portugal into a county; and the reigns of Henry and Alphonso Henriquez. 8. The erection of Portugal into a kingdom;
and

and the reigns of Alphonso I. called Henriquez, Sancho I. and Alphonso II. 9. The reigns of Sancho II. called Capel, Alphonso III. Dennis, Alphonso IV. Don Pedro, and Ferdinand, to the year 1383. 10. The interregnum. 11. The reigns of Don John I. Edward, Alphonso V. Don John II. Emanuel called the Great, Don John III. Sebastian, and Cardinal Henry, to the year 1580. 12. The reign of Philip II. King of Spain, who became King of Portugal. 13. The affairs of the Indies under the three last Kings, Sebastian, Henry, and Philip II. to the year 1640. 14. The reign of Philip IV. and the revolution in favour of the Duke of Braganza, who was proclaimed King by the name of Don John IV. 15. The consequence of this revolution, and the wars of the Portugueze against Spain, to the year 1656, 16. The reign of Alphonso VI. and the further consequence of the wars against Spain; the deposition of this Prince, and the advancement of Don Pedro his brother to the crown of Portugal; the reign of John V. and lastly, the reign of Joseph I. the present King of Portugal.

VIII. *VI. The history of Spain*, which contains the following epochs.

1. The ancient history of Spain, in part obscure and fabulous, from Japhet and Tubal to the eighth century after the birth of Christ, when the Saracens penetrated into Spain. This period

riod includes about 2862 years, and is divided into three memorable epochs; which are,

That which paſſed in Spain before the Romans:

That which was under the Romans: and

The fate of Spain after the Romans.

2. The middle hiſtory of Spain; which contains wnat paſſed from the invaſion of the Saracens and the Moors, to the time of their entire expulſion: a period that comprehends about 779 years; and during which many Barbarian and Chriſtian Kings reigned over divers provinces of Spain; and who formed the kingdoms of Caſtile, Leon, Navarre, Arragon, and Portugal, beſide that of the Saracens; and this comes down to the year of Jeſus Chriſt 1474.

3. The modern hiſtory of Spain; which begins with the reign of Ferdinand the Catholic, who united under his ſceptre all the kingdoms, provinces and colonies belonging to Spain, and formed of them one powerful monarchy. This laſt period, which includes 291 years, to the year 1765, contains the remarkable reigns of, 1. Ferdinand V. called the Catholic: 2. The ſix Kings of the houſe of Auſtria; to wit, Philip I. called the Fair, ſon of the Emperor Maximilian I. 3. Charles V. Emperor: 4. Philip II. 5. Philip III. 6. Philip IV. 7. Charles II. and three Kings of the houſe of France; that is, 8. Philip V. 9. Ferdinand VII. and 10. Charles III. In this laſt period due attention alſo ſhould be had

had to the manner in which several provinces of Africa, in Italy and the Low Countries, &c. have been united to the Spanish monarchy: and likewise the manner in which America was discovered, and reduced, under the reigns of three Spanish Kings; and many other very remarkable events.

IX. *V. The history of France.* Those historians who suppose with M. Mezerai, that the Romans first gave the name of Gaul to that large tract of land which lies between the Alps, the Pyrenees, the Mediterranean Sea, the Ocean and the Rhine, seem to have less foundation for their opinion, than they who maintain that this extensive and pleasant country was very populous in the first ages of the world, as appears by the most ancient monuments; that these people had probably a name before the existence of the Romans, and that they called themselves Galli; and, that being too numerous for their own country, they passed the Alps at the beginning of the Roman republic, and occupied a part of Italy, which was called Cisalpine Gaul; that they extended their colonies even to Asia, where they inhabited a country called Galatia, which is the name the Greeks gave to Gaul; and that other detachments of this nation advanced into Germany, beyond the Rhine. Be these matters however as they may, the history of France may be divided into several periods; the principal of which are characterized by events that are important,

portant, and proper to affift the mind of thofe who make it their ftudy.

X. The firft period comprehends the hiftory of ancient Gaul, to the time that Julius Cæfar finifhed the conqueft of that country, about forty-eight years before the birth of Chrift.

The fecond period contains the time that Gaul was under the dominion of the Romans, till the Francs entered that country, and there eftablifhed their refidence; which includes about 400 years.

The third period, which begins about the year of Jefus Chrift 412, contains the government of the Francs in Gaul; and goes only to the year 420. From that period the kingdom of France has been governed by kings that have fprung from grand houfes, which they call Races; and of which there are five that form fo many diftinct and confpicuous divifions in the hiftory of this illuftrious monarchy.

The fourth period, therefore, contains the hiftory of France, under twelve Kings of the Merovingian race, from Pharamond the firft King; that is, from the year 420 to 752; when Childeric III. after the death of Charles Martel, was depofed by the States, and Pepin elected in his ftead. This period comprehends 332 years. The firft period makes part of ancient hiftory; the fecond, third, and fourth belong to the middle age; the fifth, and all that follow, appertain to modern hiftory.

The

The fifth period contains the hiftory of thirteen Kings of the Carolovinian race, from Pepin the Short to Lewis V. called the Drone; that is, from the year 752 to 987: making 235 years.

The fixth period includes the reigns of fourteen Kings of the Capetian race, from Hugh Capet to Charles IV. called the Fair; that is, from 987 to 1328: being 341 years.

The feventh period contains the reign of twelve Kings of the Valefian race, or of the houfe of Valois; from Philip VI. of Valois to Henry III. that is, from the year 1328 to 1589: including 261 years.

The eighth period contains the reigns of four Kings, of the race or houfe of Bourbon, from Henry IV. called the Great to Lewis XV. named the Well-beloved; that is, from the year 1592 to the prefent year 1765: comprehending 163 years.

XI. After having acquired a fufficient knowledge of the kingdom of France, it is proper to be acquainted with,

VI. The hiftory of the Kings of Bourgogne; as it is not only intimately connected with that of France, but alfo throws great light on thofe of Germany, Spain, the Low Countries, &c. And here we muft carefully diftinguifh, (1.) the Kings of the firft kingdom of Bourgogne, and remember that when the Vandals, Suavians, and Alains, quitting Germany, paffed the Rhine, and
entered

entered Gaul, the Bourgognians, being amongſt them, fixed themſelves near the Rhine, and founded a kingdom that laſted 128 years; that is, from the year 406 to 534. Their government comprehended, toward the cloſe of it, the Dutchy of Bourgogne, Franch County, Dauphiny, and Savoy; under five Kings, named, 1. Gondicair: 2. Gonderic and Chilperic, who were brothers: 3. Gondebaut, Godegiſel, Chilperic, and Gondemer; likewiſe brothers: 4. Sigiſmond; and, 5. Gondemer, who was deprived of his kingdom by the ſucceſſors of Clodomire King of France; and his dominions united to thoſe of that kingdom. (2.) The Kings of BourgogneTransjurane: and we muſt here remember, that about the year 888, after the depoſition of the Emperor Charles the Groſs, Raoul or Rodolph, ſon of the younger Conrad, and grandſon of Hugh, poſſeſſed the country between Mount Jou and the Alps; that is to ſay, Savoy and Switzerland; and was crowned King of BourgogneTransjurane at St. Maurice in Valois. This kingdom laſted 145 years, under four Kings: who were, 1. Raoul: 2. Raoul II. 3. Conrad: and, 4. Raoul III. called the Drone. Conrad had united to his kingdom that of Arles; and Raoul III. having no family, left all his rich poſſeſſions to Conrod II. called the Salic: ſo that after his death this kingdom paſſed to the Emperors, who ſucceeded Conrad, and made a part of the Germanic empire. (3.) The Kings of Arles or Provence. Lewis the Stammerer,
King

King of France, dying, and leaving only princes that were quite young, Boson, brother to Queen Richilde, wife of Charles the Bald, founded the kingdom of Arles *(regnum Arelatense)* and of which he made himself King. This kingdom was surrounded by the Soana, the Rhone, and the Alps. It cannot be properly said to have existed more than 53 years; and had only two Kings, which were, Boson, who was crowned at Vienna by the Archbishop of Lions; and, 2. Lewis, son of Boson, whom Berenger took prisoner in Veronne, and whose eyes he put out. Lewis the Blind reigned, notwithstanding, forty-three years, and left a son named Charles Constantine. But as he was too young to reign, the Provencals elected Hugh King of Italy, to be King of Arles. There were great contentions between this Hugh and Raoul II. King of Bourgogne: but by the interposition of friends they were reconciled. Raoul renounced his pretensions to the kingdom of Italy; and Hugh, in return, ceded to him all he possessed in Bresse, Viennois and Provence, and even the title of King of Arles; which kingdom was therefore united to that of Bourgogne Transjurane.

XII. And here we should also study,

VII. The history of the Dukes of Lorrain, the Dukes of Normandy, Princes of Orange, &c. but we must content ourselves with the bare mention of these, without making their analysis, that we may not extend this chapter beyond its due bounds. We therefore pass directly to,

XIII. *VIII. The History of Switzerland*, or the *Thirteen Cantons*. The people that are now called *Swiss*, were anciently called *Helvetians*. About fifty six years before the birth of Christ, they made an invasion upon Gaul; but the Gauls calling the Romans to their assistance, these not only drove out the Helvetians, but brought them also entirely under the dominion of the senate and people of Rome. The capital of their country was called *Aventicum*, a small town that is now called *Wifflisburg*. The most remarkable periods in the history of the Swiss, beside the aforementioned epoch, are,

1. The time the Helvetians were in subjection to the Romans.

2. The time that the greatest part of Switzerland passed under the power of the ancient Kings of Bourgogne (see sect. XI).

3. The time when, after the extinction of the kings of Bourgogne, Switzerland became a province of France.

4. The time when Switzerland was annexed to the kingdom of Lothair (*regnum Lotharingiæ*, or Lorrain).

5. The time when it made a province of the the empire of Germany, after the dissolution of the kingdom of Lorrain.

6. The time when it made a part of the kingdom of Arles.

7. The time when it fell under the power of the emperors of Germany.

8. The time when Switzerland threw off the yoke, or when the Cantons affociated, and formed a free republic.

9. The time that has paffed fince that affociation, under the government of the thirteen free Cantons, down to the prefent day. To which may be added,

10. The hiftory of the country of the Grifons, and

11. The hiftory of the countries affociated with the Helvetic republic: and laftly,

12. The hiftory of the city and republic of Geneva.

XIV. IX. *The Hiftory of Italy*, fince the time of Charlemagne, that is, from the beginning of the ninth century to the prefent time. In this general hiftory of Italy we have to regard, in particular,

1. The hiftory of the Popes, confidered as fecular Princes and temporal Sovereigns.
2. That of the kingdom of Sardinia.
3. That of the kingdom of Naples.
4. That of the kingdom of Sicily.
5. That of the kingdom of Corfica.
6. The hiftory of the Grand Dutchy of Tufcany, or Florence.
7. The hiftories of the eight moft confiderble dutchies and principalities; which are,

a The

a The dutchy of Savoy, and the county of Piedmont.

b The dutchy of Milan.

c That of Montferrat.

d That of Mantua.

e That of Parma and Placentia.

f That of Modena.

g That of Mirandola.

h That of Monaco.

8. The history of the island of Malta; and of the religion, or order, of the Knights of St. John of Jerusalem.

9. The history of the republic of Venice.

10. That of the republic of Genoa.

11. That of the republic of Lucca. And

12. The history of the republic of St. Marino.

To which may be also added,

13. The history of the island of Cyprus, which had once its proper Kings, but has since passed under the dominion of the Ottoman empire. It would require an entire volume fully to explain all the principal periods and epochs of the particular histories of each of these governments. But, as on one hand the history of Italy is intimately connected with those of all the principal modern monarchies; and as on the other, there are many excellent histories of this country, we must refer our readers to them, as they can only expect here to find general instructions for the guidance of their studies in these matters.

XV. X. *The History of Great Britain*, or of the kingdoms of England and Scotland, and the history of Ireland. The history of England is, like most others, easily divisible into three periods, which comprehend,

1. The ancient history of England, whose original name was Albion, so called from the white chalky mountains with which the coasts of that island are surrounded. The historians begin this period with a king named Brutus, the son of Æneas Sylvius, king of the Latins, and grandson of Ascanius, the son of Æneas the Trojan. They pretend that he began to reign in Albion in the year of the world 2828. This period continued to the year 3895, and consequently till within about a century of the Christian era. This is a period of obscure and fabulous conjectures, when England was governed by kings that may be called Aborigines, or originally of the country, of which, however, they pretend to have a regular chronological list.

2. The history of the middle age of England. The Romans invaded England, under the conduct of Julius Cæsar; and though it appears they were but badly received, yet it is certain that the succeeding kings of this country paid an annual tribute to the Romans, and were obliged to suffer their prefect to reside there. This government lasted 503 years, under a long succession of kings who were natives of the country, and of whom Arthur, one of the last, was the most famous. In the fifth century the Saxons

and Angles made a defcent in England, and there eftablifhed feven fmall kingdoms; this government was diftinguifhed by the name of the Heptarchy; and lafted 369 years, and the dominion of the Saxons in England continued in all 564 years.

3. The modern hiftory of England. Soon after the death of Charlemagne, that is, in the year 801, the feven Saxon kingdoms in England were united in one, under Egbert, king of the Weft Saxons, who fubdued all the others, and reigned alone in that country. He and his fucceffors were greatly molefted by the Danes, who alfo made a defcent in this ifland, committed hoftilities, and endeavoured to eftablifh themfelves there; and which at laft they effected, in the year 1017, when Canute the Great, king of Denmark and Norway, was alfo crowned king of England. This Danifh epoch continued only 50 years, for in the year 1066, William I. called the Conqueror, duke of Normandy, landed on the Englifh coaft, drove out the Danes, and caufed himfelf to be crowned king of England. From that time England has been governed by,

Three kings of the houfe of the dukes of Normandy, from William the Conqueror to Henry I. during 70 years, and to the year 1136.

One king of the houfe of Blois, named Stephen, who reigned 19 years.

Fourteen kings of the houfe of Anjou, from Henry II. duke of Anjou, Normandy and Aquitain, to Richard III. during 331 years, down to 1485.

Three kings defcended from the earls of Richmond, from Henry VII. to Edward VI. during 68 years, and to the year 1553.

Two queens, Mary and Elizabeth, during 50 years, to 1603.

Four kings of the houfe of Stewart of Scotland, James I. Charles I. who was beheaded, Charles II. and James II. who, with the Protector Cromwell, reigned, during 85 years, to the year 1688.

One prince of Orange and Naffau, William III. crowned king of England in 1689, and died in 1702.

One queen, Ann, daughter of James II. and wife of George prince of Denmark. She died in 1714.

Three kings of the houfe of Hanover, George I. George II. and George III. during 51 years, to the prefent time.

XVI. *The Hiftory of Scotland.* The hiftorians of this country, after relating fome conjectures concerning the origin, and firft ages of the Scotch, begin their hiftory with Fergus, king of Ireland, whom the *Scoti* called from thence, and appointed their king; being no longer able to bear the horrid invafions of the Picts. Fifty-eight kings reigned after him in Scotland, during 959 years: that is to fay, from the year 411 to 1370. The laft of thefe kings was David II. who died without iffue. Robert II. fon of Walter

ter Stewart, great stewart of Scotland, and of Mary, daughter of king Robert Bruce, succeeded his uncle, and reigned twenty years. He had eleven successors of his own family; and these twelve kings of the house of Stewart bring the Scotch history down to 1603, when James VI. (and the First of England) succeeded Elizabeth queen of England, and united the two kingdoms which compose Great Britain.

The History of Ireland. If we are to believe the Irish historians, there were kings of this country more than 1500 years before the birth of Christ, and they mention one of them, named Slanius, who reigned in the year of the world 2448. They say also, that this country was divided into five governments, each of which had a king; and that, over these five kings, there was one who was supreme, and bore the title of King of Kings. There is, however, very little appearance of truth in the Irish history, till about the year of Christ 420, when a prince named Loegarius reigned in Ireland. History says that his wife and children embraced Christianity, but that he himself remained in his infidelity, and that he was killed by a clap of thunder, after having reigned 30 years. This king had forty-seven successors, who, with him, filled the throne of Ireland for 732 years; that is, to the year 1162, when this kingdom passed under the dominion of the English. The forty-eighth and last king of Ireland was named Roderic. Henry VIII. was the first English monarch who

took

took the title of king of Ireland, by virtue of an act of the parliament of Dublin.

XVII. XI. *The History of the Low Countries, and, in particular, that of the Seven United Provinces, containing,*

4 Duchies: Brabant, Limburg, Luxemburg, and Guelders.
7 Counties: Flanders, Artois, Hainault Holland, Zealand, Namur, and Zutphen.
5 Principalities, or feigniories: Friezland, Mechlin, Utrecht, Overyssel, and Groeningen.
1 Margraviate; that of Antwerp; which, together, form the

17 United Provinces, that are called the Low Countries. In the time of the Romans, the Rhine traced the limits between Gaul and Germany. That part of the Low Countries which is on the west borders of the Rhine, was named Gallia Belgica, Belgic Gaul; and that situate on the east belonged to Germany, and was called Batavia. In the fifth century, when the Franks passed into Gaul, the Low Countries remained annexed to France, under the Merovingian kings. In the partition which the Carlovinian emperor, Lewis le Debonnair, made of his dominions, the greatest part of the Low Countries falling to Lothair, made a confiderable part of the kingdom of Lorrain; and that kingdom being dissolved, the seventeen provinces above-

abovementioned were fucceffively formed. Their governors acquired great power, and at the time of the invafion of the Normans, they made themfelves independent. This hiftory therefore contains three periods. In the firft, inquiry is to be made into the origin of each dutchy, county and feigniory, till the time of their union in the fifteenth century. In the fecond, the union itfelf is to be explained, and the manner fhown in which they fell under the power, (1.) of the kings of Bourgogne, (2.) of the houfe of Auftria, and (3.) under the dominion of Spain, till the year 1564. And in the third period, it is to be explained in what manner the feven provinces of Guelderland, Holland, Zealand, Utrecht, Friezland, Overyffel and Groeningen, united themfelves, in order to throw off the Spainifh yoke, and under the conduct of the prince of Orange and Naffau, came to be declared by Spain free provinces. Laftly, is to feen the ftate of this powerful Republic down to the prefent day; together with the important eftablifhments which it has formed in the three other parts of the world, but efpecially in Afia. The hiftories of the counts of Flanders, the counts of Holland, &c. that of the princes of Orange of the houfe of Naffau, are alfo intimately connected with that of the Low Countries.

XVIII. *XII. The Hiftory of Germany.* This hiftory likewife divides itfelf into three periods, which form, the Ancient Hiftory, that of the

Middle

Middle Age, and that of Modern Germany. The firſt period comprehends the origin of the Germans, whom they ſuppoſe to be deſcended from Tuiſcon, or Teuthon, otherwiſe called Aſcanes, the ſon of Gomer, the grandſon of Japhet, and great grandſon of Noah. This chief, they ſay, began his reign in the year of the world 1812. It ſeems likely enough that theſe people took their German name Teutſche from Theuton, and that of Germans, or Germani, appears to be derived from the old German word Gerr, that is, war, and from the word Man, by which they were diſtinguiſhed as men addicted to war. Their name of Allemands came doubtleſs from Allemannus Hercules, prince of Suabia, who reigned they ſay about the year of the world 2399. It is eaſy to conceive that all this ancient hiſtory muſt be obſcure, uncertain and fabulous. The people who inhabited theſe countries knew not the uſe of letters: they tranſmitted to their poſterity the memorable actions of their founders and of the heroes of their country, by hymns and ſongs. In the Greek hiſtorians, theſe nations are always confounded under the name of Scythians, Celts, &c. and it is impoſſible to diſtinguiſh them. The firſt knowledge we have of them muſt therefore be from the Romans; who thought them worth the trouble of conquering, and had connexions with them: and conſequently all that we can learn of them muſt be drawn from Strabo, Ptolemy, Cæſar and Tacitus: and theſe authors did not even underſtand

the

the language in which those historic hymns were sung. It appears by these writers that the ancient Germans were mere barbarians. Among that dark ignorance and that ferocity with which they were surrounded, there were however to be seen some sparks of virtue, valour, art and knowledge. Tacitus says, for example, that they were much addicted to drinking; and this implies that they had the art of making wine, or some other strong liquor: that author, indeed, expresly says, that they brewed beer (cerevisia). He says also, that they trafficked with the Romans, and sold them, among other things, amber, which they gathered on the borders of the Baltic Sea, and named Glæse. All this supposes some exertion of industry. This first period comes down to the birth of Christ.

XIX. The middle age comprehends the revolutions in Germany from the commencement of the Christian era to the time of Charlemagne, including eight centuries. It is in this second period that we find, (1.) The accounts of the wars that the Germans sustained against the Romans, who were never able completely to subdue them. (2.) The particular enumeration of the different nations that then inhabited Germany. (3.) The progress of each of these people; their state during the decline of the Roman empire, and the manner in which each of them insensibly recovered their liberty. It is a matter well worthy of remark, that during all the middle

dle age, the Germans remained ignorant of the art of writing, and that Charlemagne was the first who taught them the use of letters. All therefore that has been wrote of the history of the Germans during the middle age, was either by foreigners, or by monks, and others equally ignorant, after the eight century. It is sufficiently manifest what regard ought to be paid to these. The greatest inconvenience is, that we cannot form a just and distinct idea of the state of the German nations before Charlemagne. That they had chiefs is certain, but the same chief of a nation is sometimes named rex, and sometimes dux, princeps, margravio, or comes, and sometimes still different from any of these. All the *Scriptores rerum Germanicarum* of the middle age, are but so many troubled and confused sources: the business here, however, is to know what has been wrote, rather than what has really happened.

XX. The modern history of Germany begins with Charlemagne, and comes down to Francis I. that is to say, down to the present time. The history of these emperors being already included in that of the empire, it only remains in the third period of the history of Germany, to consider, (1.) The particular history of the twelve grand sovereign houses of Germany, which are those of Austria, Brandenburg, Bavaria, Baden, Brunswick, Anhalt, Hesse, Holstein, Mecklenburg, Nassau, Saxony, and Wirtemberg. (2.) That

That of the states and countries which these houses possessed. (3.) That of the archbishopricks, bishopricks, abbies, military orders, &c. (4.) That of the free cities and those that hold immediately of the empire, &c. To which may be added divers historical matters, as (5.) an inquiry into the origin of electors, and in what manner the chiefs of divers Germanic nations recovered their liberty by the right of postliminy, after the extinction of the Carolovinian house. (6.) The particular history of the Anseatic league. (7.) That of the war of thirty years; and numberless other particulars which relate to the Modern History of Germany. The history of the house of Austria, and that of Brandenburg, merit a more particular study; because the heads of these august houses are at this day elevated to the first rank among the sovereigns of Europe.

XXI. *XIII. The History of the Kings of Bohemia.* This country, situate on the borders of the Elbe, was anciently inhabited by the Sclavi, whom they named Behemanns or Behaims, for it is not more than 200 years since they called Bohemia, Behaigna. It was originally governed by dukes, the first of whom, named Zicco, conducted, with his brother Lecho, in the year 550, a powerful colony into this country, uninhabited, and almost covered with forests. From his time there have been twenty-two dukes in Bohemia, the course of 536 years, down to the year 1086. The last of these dukes was named Uladislaus II.

six

six years before whose death Bohemia was honoured with the title of a kingdom by Henry IV. and Uladiflaus reigned in quality of king. There were twenty kings his succeffors to the year 1307, when Henry duke of Carinthia, and king of Bohemia, was depofed. From that time feventeen kings and one queen of the houfe of Luxemburg and that of Auftria, have reigned in Bohemia; the firft of whom was John of Luxemburg, fon of the emperor Henry VII. At this time Bohemia makes part of the hereditary eftates of the houfe of Auftria.

XXII. *XIV. The Hiftory of the Kings of Hungary.* That country which is now named Hungary, was formerly called Pannonia. The Huns, a Gothic people, eftablifhed themfelves there, and gave the country their name. Attila made his city the capital of Sicambia, and gave it the name of Buda, which was that of his brother. The Huns gave themfelves very little trouble about writing their hiftory. We know that there were at firft dukes in Hungary, and that in the year 1000, it was erected into a kingdom in favour of Stephen, called the Saint. That king had fixty-fix fucceffors, down to the prefent day; and Hungary alfo makes part of the hereditary dominions of the auguft houfe of Auftria.

XXIII. *XV. The Hiftory of Poland.* This country was anciently named Sarmatia, and its inhabitants Sarmates. We have only a confufed account,

account, that this country was originally governed by dukes or princes, of whom there were twelve from Craco and his brother Lechus I. during 450 years: that is, from the year of Chrift 550, to 999: that on the laft named year Poland was erected into a kingdom by the emperor Otho III. that it had afterwards four Polifh kings who reigned during eighty-two years: that the laft of thefe kings, Boleflaus II. called the Cruel, occafioned this country, by his bad conduct, to lofe the title of kingdom; and that it was governed from 1081 to 1370, by twelve princes, among whom was the renowned Piaft: that in the year 1370 it refumed the rank of a kingdom; and that it has fince had eighteen elective kings, who have been chofen as well from foreign houfes, as from the Piafts, or original families of the country: that the firft of thefe elective kings was Lewis king of Hungary, and the laft Auguftus II. elector of Saxony; and that this prince dying in the year 1763, the ftates of Poland have placed on the throne Staniflaus II. of the family of Poniatowfky, a prince in every fenfe worthy to wear that crown. The Hiftory of *Lithuania* is comprehended under that of Poland. The hiftory of *Pruffia* is likewife included, in part, under that of Poland, in part under thofe of the orders of the Teutonic knights, and the knights Templars, and in part under that of the houfe of Brandenburg. The hiftory of *Finland, Livonia, Eftheria* and *Courland,*

is

is comprised under those of Sweden, Russia and Poland.

XXIV. *XVI. The History of the kingdom of Denmark.* If we regard what is said, and still more, what they have not scrupled to write, we must begin this history with Gomer II. great-grandson of Japhet, who was the first that inhabited Cimbrica Chersonesus, or Cimbria, 1800 years after the creation of the world, 193 years after the deluge, and 2098 before Christ. This country, they say, was governed, at first, by eleven successive judges, the first of whom was cotemporary with Abraham: that in the year of the world 2910, and 1058 before Christ, Dan founded the kingdom of Denmark, and called it after his name: that it had twenty-six Kings, all of whose names they know, and their principal actions, to the time of Dan III. who began to reign 141 years before the common era: that 110 years before this epoch, there was a great migration of the Cimbri and Teutoni, who penetrated into Italy; but were there almost entirely extirpated by the Romans; and at this time it is that the ancient history of Denmark ends, that is, about seventy-four years before the birth of Christ. That of the middle age begins with *Fridlef* I. surnamed the *expeditious*, who was the twenty seventh king; and continues till Sigefroi, whose reign began about the year 760, and ended with the eighth century. This age comprehends a succession of
thirty-

thirty one kings. But they whose design it is to know what has really happened, will give but little credit to all these relations, as they will not believe it possible that a nation, which knew not the use of letters till a long time after the reign of Charlemagne, should be able to trace their origin to the time of the deluge; or that they could, by any monuments whatever, be able to deduce their history from that period, without interruption, down to modern times, that is to the ninth century: they will therefore be persuaded that all those ancient histories and chronicles, in verse and prose, on which their authorities are founded, are nothing more than a mass of fables, written by impostors and visionaries two or three thousand years after the facts are supposed to have happened, and consequently that they knew no more of the matter than we do. Without making any further inquiry therefore into these relations, we shall say, that the modern history of Denmark, which begins about the year 800 with Goteric, Godfrey, or Gotilae, is more clear and less uncertain. It includes the reigns of fifty-five kings, during 965 years, that is, from the year 801 to the present time. So that the historians count one hundred and thirteen kings who have reigned in Denmark, from Dan I. to Frederic V. who now so worthily fills that throne. The introduction of Christianity into Denmark, under Eric I. and the sixty-seventh king, about the year 850, falls in this last age, which also abounds with remarkable events.

events. The history of *Norway* is included in that of Denmark; as are those of the islands of *Iceland* and of *Greenland*, if any such there are.

XXV. XVII. *The History of Sweden.* The ancient history of this kingdom is altogether as fabulous as that of Denmark. All its first accounts consist of relations, songs and legends, of the allegoric traditions of ancient priests or poets. On these authorities they suppose that *Magog*, the son of Japhet, and grandson of Noah, was the origin from whom descended the Scythians and Goths. Magog left five sons. From one of them, named *Suenon*, they say the Swedes are descended: from *Gog* or *Gethar*, they suppose the *Goths* or *Getæ* derive their original. *Ubbon* succeeded his brother *Suenon*, and built, they say, the city of Upsal. They here make a succession of twenty-five fabulous kings, from *Magog* to *Bothavill*, and which comes down to the birth of Christ.

The middle age begins with the reign of *Alaric*, and after him of *Eric* II. and continues to *Ingo* II. or *Ingel*, and lasts about 800 years; including the reigns of thirty-five kings, whose history is scarce more certain than that of the former.

The modern history begins with the kings Charles and Biorn. About the year 831, the emperor Lewis the Debonnair sent Ansgairus, bishop of Bremen and Hamburgh, into Sweden, to preach the gospel, and Christianity was received

ceived into that kingdom. From Charles to Suercher II. that is, to the year 1140, there were twenty-two other kings who reigned in Sweden, during a space of 302 years: and from Eric IX. called the Saint, who succeeded Suercher II. there is a succession of forty kings, who have filled the Swedish throne, down to the present day, during 615 years. This long series of 122 kings is very gloriously terminated by the reigning monarch Frederic Adolphus, a prince of the house of Holstein. The history of *Lapland*, as far as any history can be found of that country, is entirely included in that of Sweden.

XXVI. *XVIII. The History of Russia.* All that we can say of the ancient history of Russia is, that this country was the hive from whence that innumerable swarm of people issued, which overran all Europe, and part of Asia and Africa. It would be a vain and frivolous enterprise to endeavour to investigate either the ancient history, or that of the middle age, of this immense country, as no written accounts of them have come to our knowledge. The foundation of this vast and formidable empire, which was civilized by the labours of Peter the Great, and of those princesses who have borne the sceptre after him, and who at this day make so conspicuous a figure on the theatre of the world, was not laid till the ninth century of the Christian era. The same origin is usually ascribed to the inhabitants of this country as to those of Poland and Bohemia.

The language called Sclavonian, which they all speak, but in different dialects, seems to confirm this conjecture. All the colonies of the great nation of Sarmatia took, in the sixth century, the name of *Sclavi*; by which they meant to express that they were pursuers of glory, for such is the import of that word in the Sclavonian language. That of *Ruſſia*, or *Roſſaia*, implies a people that are difperfed; and this etymology is confirmed by Procopius, a Greek hiftorian of the fixth century. The ftory of one Ruſſus, the brother of Lexkhus and Czekhus, is an invention of modern writers among the Illyrians or Dalmatians.

In the years 861 and 862, the inhabitants of Ruſſia chofe for their governors three Varegean princes, Rurika, who firſt fixed his refidence at Ladoga; Sineus, at Bielo Ofero, and Truvera at Ifborka. The two laſt died without children in the fpace of two years. By that means Rurika became fole fovereign of Ruſſia; and having augmented the city of Novogrod, which had been lately built, he there eftablifhed his refidence. In 878, this great prince, Rurika, died, and left Igora, his fon, under the tutelage of Olegha, his uncle; he governed Ruſſia thirty-five years. When Igora came to age of maturity, he efpoufed a young maiden of Plefkow, named Olpha. This prince was maſſacred by the Drevelians about the year 945, and Suetoflava, his fon, reigned in his ftead, under the tutelage of his mother Olpha, with whom he always

ways lived in harmony. This princefs embraced
the Chriftian religion at Conftantinople, and was
baptized by the name of Helena. Her fon,
however, did not follow her example. Jarapolka
reigned after his father Suetoflava, from the year
972 to 988. His brother Vladimire, or Wolo-
dimir, called Bafil, fucceeded him, and embraced
the Chriftian religion according to the rites of
the Greek church. From this point, the hiftory
of Ruffia becomes more luminous: for, with
the doctrine of Chriftianity, the ufe of writing
was introduced among this people. From Wo-
lodimir or Bafil, to Bafil V. during 546 years,
that is, from 988 to 1534, we find a fucceffion
of thirty-five fovereigns, who reigned in Ruffia
under the title of Great Dukes. John IV. or
Iwan Bafilowitz, the fon and fucceffor of the laft
Bafil, took the title of Tzar, or Czar, which his
fucceffors have continued to bear, and which, in
the Sclavonian tongue, properly fignifies a king.
He alfo joined to his titles that of Povelitela and
Samodertza, confervator or fovereign of all the
Ruffias. Foreigners call this prince the Tyrant,
but the Ruffians name him the Severe. He had
fix fucceffors, who contented themfelves with
the title of czar, till the year 1613, when Michael
Fedorowitz, of the houfe of Romanova, mounted
the throne, and took the title of czar, emperor,
and autocrator (or fovereign confervator) of all
the Ruffias. This title of emperor is no longer
contefted with thefe powerful monarchs. From
Michael Federowitz there were three emperors

of

of Russia, Alexis his son, Teodore or Theodore, and Ivan, or John V. to the year 1696, when Peter I. surnamed the Great, came to the throne: a monarch who made the most astonishing, and at the same time the most happy efforts, toward the civilizing of the Russian nation. This great man died in 1725; and the throne of Russia has been since filled by Catherine, the widow of the emperor Peter; by Peter II. his grandson; by Ann, the daughter of John V. by John VI. grandson of John V. by Elizabeth Petrowna, daughter of Peter the Great; by Peter III. grandson of Peter, and lastly, by the august Catherine II. now reigning.

XXVII. If the Cossacks, Calmacs, the inhabitants of Siberia and the Ukraine, the Samoeids, &c. have any history, it must be comprehended under that of Russia. It will not be expected that we should lose ourselves in these desarts. We must here say a few words, however, concerning the *Tartars*. Tartary in Asia, which they call *Great Tartary*, is an immense country, that is as imperfectly known to geographers, as the succession of its sovereigns is to historians and chronologists. It was these Tartars, however, that in the year 1280 made themselves masters of China; and it was then that the family named Ivan began to reign there. There were nine Tartarian emperors of that house, which lasted 89 years. In 1369 the Tartars were drove out of China; but in 1645 they re-entered, eered, under the command of their cham, or king, Xun Chi, whom they named the Grand Kam; again subdued the empire of China, and the family of that Tartar prince continues to reign there till this day. *Little Tartary* comprehends all that country which is between the Tanais and Borysthenes. It is far from being precisely known what time the Tartars made themselves masters of this country, for the opinions of historians differ widely concerning this matter. That which appears the most probable is, that the dukes of Lithuania having subdued the Tartars, they sent princes of their nation to rule in this country. The last of these princes was named Aczkirei, from whom came the race of Girei, and all those pretended emperors who have reigned in Little Tartary since the year 1452. About the middle of the sixteenth century, Selim, emperor of the Turks, subdued Crim Tartary, and took the fortress of Cafa: and from that time the Kam of the Tartars has been chosen by the Ottoman Porte: sometimes indeed they have suffered the eldest son of the Kam to succeed him, and at other times they have not. We have seen, moreover, a Kam called to Constantinople to give account of his conduct, and sent into banishment. We are the better pleased with this opportunity of mentioning the Tartars, as their history leads us to make three reflections. The first is, that we cannot conceive from whence the illustrious author of the Persian Letters could learn that the Tartars conquered

quered almoſt the whole world. For if that were the caſe, it certainly was not ſince that people who inhabit Great and Little Tartary have borne the name of Tartars. Perhaps he thereby means the Scythians or Celts, or ſome other ancient and warlike people. The ſecond obſervation is, that if modern hiſtorians know ſo little of the revolutions and actions of a nation that is now exiſting, and before our eyes, but who are ferocious, uncivilized, and have no writers among them, what are we to think of the ancient hiſtories of all nations, and eſpecially thoſe of the north, who, for thouſands of years, were in the ſame circumſtances, and were ignorant of the uſe of characters, and the art of writing? And laſtly, that thoſe philoſophers deceive themſelves, who imagine that a nation becomes more formidable by being ignorant of arts and ſciences. The example of the Tartars ſufficiently proves the contrary, and ſhows that a people may be numerous, brave and warlike, and yet not able to form themſelves into a body as a nation, and ſtill leſs able long to ſupport themſelves, if they do not become civilized, and cultivate the arts and ſciences. The Goths and Vandals proved this truth formerly. What remains of thoſe people are there now upon the earth? If any of them can be ſaid ſtill to exiſt, they are become civilized; for the children of the Goths and Vandals that were born among poliſhed nations acquired the manners of thoſe people from their infancy.

XXVIII.

XXVIII. To render the study of modern history complete, we must transport ourselves into the other three parts of the world, and we live in an age when we can make those journies without going out of our closets. In Asia we have to consider, beside the empire of the Turks, which we have already mentioned,
1. The modern history of *Persia*.
2. The like history of the *Moguls*, or emperors of *Indostan*.
3. That of the kingdoms of *Pegu*, *Ava*, and *Arracan*, or of those countries which the ancients comprehended under the name of the Indies beyond the Ganges.
4. The history of the kingdoms of *Siam*, *Laos*, and *Tonquin*.
5. That of the kingdom of *Bengal*, and its nabobs.
6. The modern history of *China*.
7. The history of *Japan*.
8. The history of the kingdom of *Java*.
9. That of *Ceylon*, formerly called *Taprobane*.
10. The histories of the other *large islands of the Indian and Oriental sea*.

XXIX. In Africa, we have also to learn, beside what is under the immediate dominion of the Ottoman empire,
1. The history of *Abyssinia*.
2. That of *Tunis* and *Tripoly*.
3. That of *Algiers*.

4. That

4. That of the kingdom of *Morocco*, under which are comprehended those of *Fez*, *Taffilet*, *Tetuan*, *Sus*, and others.

5. The history of the other nations of Africa, as far as they can be, and are worthy to be known.

XXX. In America, we have lastly to consider,

1. The history of the *Canary Islands*.
2. The history of the *discovery of America*, and the progressive manner in which we have become acquainted with all its various countries, as well islands as continent.
3. The history of the *partition of America* among the European powers; to which may be added,
4. The history of *Mexico*, and
5. That of *Peru* and its *Yncas*.

XXXI. Such in general is the system of what is called the universal history of the world; of the ancient and middle ages, and of modern times. It must be confessed that the labours of the learned have, in this science, surpassed all that we could expect, and all that the capacity and assiduity of the human mind seemed capable of producing. There are now, in almost all languages, universal and particular histories that are highly excellent; where the most learned researches are united with the most sagacious reflections,

flections, and where a regular and conspicuous narration is ornamented with all those graces of which the historic style is susceptible. There are in the universities able professors, who make courses in history that are highly instructive: and there are historical bibliotheques which furnish us with the knowledge of the best authors in every species of history. They therefore who are desirous of applying to this science, cannot want for guides, or instructions; and we may add, that, in this age, the useful and the agreeable will be found united in the study of history.

CHAP. VIII.

THE ECCLESIASTICAL HISTORY OF ALL THE Principal Nations of the Earth.

AFTER having treated on profane or civil history, we naturally proceed to *Sacred*, or *Ecclesiastical History*. From the first hour that

that man comes into the world, he seeks after happiness. The milk which nature has given to the mother, renders the child content, and consequently happy: it desires nothing more. But as its age advances, and its ideas increase, it seeks after sports, pleasures, and fortune. At length man discovers that there is a future existence, and a Supreme Being, who is the creator and preserver of this world, and the dispenser of happiness or misery in the world that is to come. The first human beings perceived therefore that it was of importance to render the Divinity propitious to them; but to obtain that end, they made use of means that were as weak and imperfect as were their understandings. Having nothing but sensible objects before their eyes, they could form only corporeal ideas, and these ideas they applied to the Supreme Being, to whom nothing corporeal can belong.

II. By groping continually in that darkness, without the guidance of revelation or philosophy, after the right way of obtaining the knowledge of God, and the manner in which he ought to be worshipped, they could not but wander and deceive themselves, as well with respect to the one as the other of these objects. The knowledge of God, and of the manner in which he is to be worshipped, forms what we call religion. The design of this chapter, therefore, is to inform our readers what have been the principal religions, that men have invented and followed,

from

from the creation of the world; and the following chapter will contain the history of Christianity, or of the church of Christ in particular.

III. Adam, and the first patriarchs after him, followed, doubtless, the religion of nature; the lights of reason, enforced by those which God had vouchsafed them in Paradise, and in the succeeding ages; as we find in the book of Genesis, wrote by Moses. But this worship, so pure in itself, seems to have been sometimes corrupted by a propensity to idolatry, which infected mankind from the beginning of the world. The sacrifices of animals, and even of innocent men, are not certainly according to the religion of nature, but have a near relation to paganism. For all sacrifices are diametrically repugnant to the religion of nature, as no man can possibly prove, by the light of reason, that the Supreme Being, all-wise and good, can find pleasure in the slaughter of his creatures, and what is more, of mankind; whom his wisdom has created, and whom his goodness supports. The little houshold gods of Laban, the father-in-law of Abraham, clearly prove that idolatry reigned in the first ages of the world. Moses purged the worship of the Hebrews entirely from it; it was he who, by the express order of God, established the true principles of religion among the children of Israel; their dogmas and their religious ceremonies. We are therefore here to consider:

IV. (1.)

IV. (1.) *Paganism.* We have already given an ample description of this religion in the second chapter, on mythology; and we shall only add here, that paganism in general has at all times had various sects, and that even when it possessed almost the whole earth, each people had their different gods, idols, and religious worship; at least with regard to exterior matters. The pagan religion of the Egyptians, for example, was not the same with that which was professed by the Greeks, and theirs differed likewise from that of the Romans, who multiplied their demigods and temples to an endless number. It is a singular circumstance, and well worthy of remark, that, even in modern times, whenever a nation or troop of mankind are discovered in any part of the earth, they are always found to be pagans. Whence comes it that mankind have naturally so universal a propensity to idolatry, and so little to philosophy and the principles of Christianity? Whencesoever that be, paganism was destroyed in the reign of Theodosius the Great, at the close of the fourth century of the Christian era; and the ruins of it which are to be found in Asia, Africa, and America, are degenerated into an absurd idolatry; and always attended by ferocity, ignorance, and barbarity. That large work, of " the religious ce- " remonies and customs of all nations, repre- " sented by figures designed by Bernard Picart, " with an historic explanation, &c." and especially those volumes which treat of the idolatrous nations,

nations, is very inftructive, and throws great light on thefe objects.

V. (2.) *The ancient religion of the Chinefe:* which is but little known to us. We know that they adored the heavens, under the name of *Thien*; and that they had in their devotion fome mixture of that of the Jews, though we know not from whence they had it. There is a very ancient tradition among the Orientals, that there are a great number of Jews in China, and that God having opened a paffage, they went thither in the time of Jofhua. However that be, it is certain that a large portion of idolatry, fome principles of natural religion, and of that of the Hebrews, formed the religion of the ancient Chinefe. But about 550 years before the birth of Chrift, that is about the year of the world 3450, the renowned Confucius was born in the kingdom of *Lu*, which is the province that is now called Xantung. This philofopher was of an illuftrious family, that defcended from the emperor Ti-Ye, of the fecond race. He began by profeffing philofophy, and ended by inventing a new fyftem of religion and politics. His reputation acquired him more than three thoufand difciples, among whom there were feventy-two that fignalized themfelves, and are ftill held in great veneration by the Chinefe. Confucius divided his doctrine into four parts, and his difciples into a like number of claffes. The firft were thofe who applied themfelves to
the

the study of virtue: the second, such as applied themselves to the arts of reasoning and eloquence: the third, they who studied the art of government, and the duties of magistrates; and the fourth, those who applied themselves to the doctrines of morality. The four books that are attributed to Confucius are considered by the Chinese, as of the highest authority. The first is intitled *Ta-Kio,* or the grand science. There is only the first chapter of that book that properly belongs to Confucius. The rest of it, as well as the second, called *Chung-Yung,* or the medium of virtue; the third named *Lungya,* or the conferences; and the fourth which is a collection of conversations: all these books are the works of his disciples. Though it is said, in all these books, 1. That it is the heavens or virtue that holds the place of the Supreme Being, yet 2. They direct superstitious worship and sacrifices to others than that Being, and 3. They promise no other recompence or happiness than that of this life. In the modern religion of the Chinese, which is founded on the doctrine and writings of Confucius, there are three sects, *the Learned, the Idolaters, and the Sorcerers.* The first is that of the emperor and nobles, who sacrifice to the stars: the second pay their adorations and build temples to idols; and both of them render a religious worship to Confucius, to philosophers, to kings and their ancestors. The third sect worship demons and practise magic. The Chinese

nese priests are named *Mandarins*; and apply themselves to religious affairs, to philosophy and government. There are many temples and convents in all parts of China. The idols of the Chinese are called *Pagods* or *Chines*. The latter are made in the shape of figured pyramids; and are held in great awe by the vulgar. When they purchase a slave, they bring him before one of these chines, and after making an offering of rice, or other matter, they entreat the idol, that the slave, if he should fly from his master, may be devoured by tigers and serpents: and this the slaves fear to so great a degree, that they never dare to leave their masters, whatever may be the treatment they receive. Idolatry therefore is very manifest in the religion of the modern Chinese, but Confucius is not to be blamed for this error; for in the first chapter of the book Ta Kio, which is the only one that he wrote, there is no trace of it to be found. All the rest is the work of his disciples, a class of men who constantly enlarge, decorate, and disfigure the doctrines of their masters. Notwithstanding all the absurdities which we discover in the religion of the modern Chinese, that people have lived, for 2000 years past, in peace and tranquillity under its shadow, and have derived from it an exterior happiness.

VI. (3.) *The Religion of the Magi.* The word Magus in the ancient Persian is nearly synonymous with that of sage or wise man: and this name was given to those philosophers who taught morality and natural theology, founded on the adoration and worship of a Divinity, as Arnobius has remarked. This natural religion, however, was not either very pure or very rational; for the magi laid down two imaginary principles, which were, that *light* was the source of good, and *darkness* the origin of evil. These philosophers, however, were in high estimation with the kings of Persia, who acknowledged their wisdom, and honoured them with the name of Sages; frequently consulted them in the affairs of government, and charged them with all that regarded the religion and policy of their kingdoms; so that they were at once priests, politicians and philosophers. It is easy to conceive what importance this triple employ gave them in their country; and the more, as by the study of natural philosophy these magi were enabled to predict appearances in nature, and sometimes to perform operations that appeared supernatural to the people, and which these subtle priests caused to pass for conjurations, prodigies and miracles. When Cambyses had determined to carry the war into Egypt, he appointed one of these, named Patizithes, governor in his abscence. But that minister attempting to place his brother Smerdis on the throne, in the room of the son of Cyrus, whom Cambyses had slain, the principal satrapes or nobles, perceiving his fraduluent design, massacred, at once, him and all the rest of the magi. From the time of this catastrophe, the

sect

sect of the magi fell into disrepute; but, some years after, they were restored to authority, and at the same time reformed by Zoroaster. They, who in succeeding times made a profession of sorcery, took the name of magi, and from thence a bad signification was annexed to that title, and from thence also is derived the word magician. These magi spread themselves over all the East, and even in Egypt, where we find them in the time of Moses. The priests of the sect of magi in Persia were all of the same tribe; and they rarely communicated their science to any but those of the royal family, who from thence were regarded as belonging to the sacerdotal tribe. These priests were divided into three orders; the common clergy, the superiors, and the archimagus, or head of their religion. The temples were in like manner of three orders. The archimagus held his residence in the principal temple, and the whole sect thought themselves obliged, once in their lives, to go thither on a pilgrimage. The business of these priests was to read the offices of each day in their liturgy, and at certain fixed and solemn times to explain to the people different parts of their sacred books. There were no altars in these temples; but they preserved sacred fires, in lamps, before which they performed their adorations. This people were in great dread of spectres or apparitions.

VII. (4.) *Zoroaster*, whom the Persians called Zerbusht, was, according to oriental writers, a great philosopher, who lived at the time that Darius, the son of Hystaspes, filled the throne of Persia. He was perfectly acquainted with all the oriental sciences, and much versed in the religion of the Jews. He did not found a new religion, but undertook to reform that of the magi, which for many centuries had been the prevailing religion among the Medes and Persians. He established the doctrine of *a first principle*, or *Supreme Being*. He taught that fire was the symbol of the presence of the Divinity, and that God had established his throne in the sun. He shut himself up, for a long time, in a cavern of Media, where he composed the book of his *Revelations*. A short time after, he went into Bactriana, and Persia, and there caused his doctrine to be received. From thence he passed into India, in order to learn the sciences of the Brachmans; and having acquired all they knew of physics and metaphysics, he returned into Persia, and communicated his knowledge to the magi; who from that time were held in high esteem. Zoroaster, repairing to the court of Darius at Susa, presented that monarch the book he had composed, bound in twelve volumes, each of which contained a hundred skins reduced into vellum, on which it was the custom of the Persians to write. This book was intitled *Zendavesta*, and by contraction *Zend*; a word that signifies *the fire lighter*. The king, his courtiers, and the nobles of the land, embraced magianism, thus reformed by Zoroaster; maugre the

efforts

efforts of the chiefs among the Sabeans: and this religion continued to prevail in Persia till the time it was superseded by the doctrine of Mahomet. Its morality was pure, except that it permitted incest. With regard to the worship of this religion, it was simple: philosophy and policy appear to have been there artfully united. They say that Zoroaster, who retired to Balch with the quality of archimagus, was there slain by Argasp, king of the Scythians, and his temples demolished. The disciples of Zoroaster, who still remain in Persia, are called by the Mahometans Gaures or infidels.

VIII. (5.) *Judaism.* Moses who lived about the year of the world 2550, near 500 years before Homer, and 900 years before the philosopher Thales, was the first who gave a form to the religion of the Jews, reduced it into a system, and prescribed them a law as he had received it from God. This law is contained in the pentateuch of Moses, which comprehends the books of Genesis, Exodus, Leviticus, Numbers and Deuteronomy, which are in the hands of all Christians in every part of the earth. Leviticus properly contains the law, the sacrifices and ceremonies of the Jews, and Duteronomy serves as a recapitulation or abridgement of the law. The ten commandments form a kind of summary of all the fundamental laws that God prescribed by Moses to the people of Israel. All these laws are either religious and doctrinal, and relate to

the

the dogmas or essence of the Jewish religion; or ceremonial, and regard its rites and ceremonies; or civil and political, and regulate the constitution of the Judaic republic, or their police, and prescribe in a particular manner such rules as were proper to be observed by that intemperate and leperous people: or lastly moral, and served to regulate the manners and consciences of the Hebrews, by exciting them to virtue. These Divine laws, however, they did not always practise; for, when we read their history, we find, that a more profligate, cruel, covetous and deceitful people scarce ever existed upon the earth. To all their other vices they joined a strong superstition. Their Talmud, which is a sort of dogmatic catechism, or amplification of the law of Moses, is the quintessence of absurdity; and the writings of their rabbies and cabalists contain the most complete collection of insipid whims that it is possibe for fanaticism to conceive. Since the promulgation of the Christian religion, the Jews have been dispersed over the face of the earth, and no where united in a national body.

IX. (6.) *Christianity* arose, about the year of the world 4000, out of Judaism, at the time that it was become greatly corrupted. Jesus Christ appeared upon the earth, taught a doctrine that is perfectly divine, and founded a church that has spread itself into all the four parts of the world

world; and of which we shall give a brief history in the following chapter.

X. (7.) *Mahometanism.* Mahomet, called the prophet, was an artful impostor, and of his kind, perhaps the greatest man that ever appeared upon the earth. He was born the 5th of May in the year 570 of the Christian era. His father, who was an Arab and a Pagan, was called Abdalla, and his mother, who was a Jew, was named Emina, and they were both of the dregs of the people. It would require a volume to show by what address, what subtle genius, what extensive schemes, what resources, by what a bold and daring spirit, he became enabled to produce a new religion, and to establish it in Asia, Africa, and even in some countries of Europe; by bearing in one hand the Coran, and in the other the sword; and by succeeding equally well, as conqueror, legislator and prophet. The Mahometans acknowledge that Judaism and Christianity are true religions; but that they no longer contain any certain principles, because their holy books have been corrupted. They say that God communicated himself to his prophet Mahomet, by the angel Gabriel, for the space of twenty-three years; and gave him a certain number of written sheets, from whence he composed the book called the Coran or Alcoran. M. du Ryer has translated this Alcoran into French; and M. Prideaux and count Boulainvilliers have each of them wrote the life of Mahomet. The principal

principal dogmas of the Mahometan religion are, the unity of God; that there is no other God but God, and that he is one: that Mahomet was sent from God, and was his prophet, and that this last truth has been confirmed by numberless miracles (which always appear ridiculous to those that are not of the same religion). The Mahometans have also their saints to whom they likewise attribute miracles, but inferior to those of their prophet. They acknowledge, moreover, that there are angels, who are the ministers of the commands of God: they believe in a general refurrection of the dead; in a day of judgment; in a hell; and paradife, whose delights are painted by the Coran in the most pleasing figures, and with the most glowing colours. It is represented as a delicious garden, watered by fountains and rivers of milk, of wine and honey, and adorned with trees that are for ever green, and that bear apples whose kernels turn into women, who constantly preserve their youth, their beauty and virginity, and are of so sweet a nature, that if one of them were to spit into the ocean, all its salt waters would become immediately fresh. The Mussulmans likewise believe in predestination; and say that no good or evil arises but by the ordinance of God: and if they are asked, why God has created the wicked? they reply, that it is not for us to search too curiously into the secrets of the Almighty; that what appears good in the eyes of man, may be found evil before God, and that good which we call
evil,

evil. They admit of polygamy, or a plurality of wives, and forbid the use of wine and other strong liquors. They have adopted the Jewish custom of circumcision. Their morality consists in doing good and avoiding evil. They hope for the mercy of God, and the forgiveness of sins, and recommend, in a particular manner, prayers, and ablutions or the use of baths, that is corporeal purity. Christian divines have frequently attributed to the Mahometans errors which they do not profess: it must be acknowledged, at the same time, that the Coran, notwithstanding all that we there find, which is sagacious and even sublime, abounds with absurdities and such idle tales as are offensive to common sense. We ought not, however, always to attribute these to Mahomet, for they are frequently the produce of his commentators, and of the enthusiastic spirit of the oriental nations.

XI. The Mussulmans are at this day divided into two principal sects, and who are even mortal enemies to each other. The Persians glory in being the followers of Ali, and wear a red turban. The Turks, on the contrary, hold the memory of Ali in contempt, following the sect of Omar, and wear a white turban. There are many other sects among the Mahometans, of whom they count even sixty-seven. All these sects, however, occasion no schism, but agree in their fundamental dogmas; pray, give alms, make

make the pilgrimage to Mecca, and obferve the faft of Ramadan.

XII. It remains to fpeak of certain religions, of which, though not generally received, but are or have been lefs diffufed among mankind than the preceding, we ought not to be ignorant at leaft of the names, if we would attain a complete idea of the various worfhips and fuperftitions that have reigned among the human race from its firft exiftence. Such are,

(8.) *The Religion of the Bramins*, or the inhabitants of Tonquin, between China and India. Brama is their principal god, and adored by the followers of Confucius. They have likewife three other divinities, who are Raumu, Betolo, and Ramonu, and one goddefs, who is called Satibana. Befide which they facrifice to the feven planets as divinities. The people, but efpecially the priefts of this fect, are named Bramens, Bramins, or Bramines, and thofe names are formed from the word Brachmanes, by which the Greeks and Latins denoted the Indian philofophers. They believed in the immortality of the foul, but they added to that belief the metempfychofis, or tranfmigration of the foul from one body to another.

(9.) *The Religion of the People of Barantola*, in Southern Tartary, in Afia. This kingdom is governed by two fovereigns. The firft, who is charged with the political government, is named Deva; the other, who lives retired, is not only adored

adored by the inhabitants of the country as a divinity, but alfo by the other kings of Tartary, who fend him prefents. This falfe god is called Grand Lama, that is to fay, Great Prieft; or Lama of Lamas, Prieft of Priefts. He is believed to be eternal; and the other lamas ferve him, and report his oracles. He is fhown in a fecret apartment of his palace, illuminated with an infinite number of lamps; he appears covered with gold and diamonds, and is feated on an eminence adorned with rich tapeftry, and fits with his legs croffed. He is fo much refpected by the Tartars, that they, who by rich prefents can obtain a part of the excrements of the grand lama, efteem themfelves extremely happy, and carry them about their necks in a gold box, in the manner of a relick.

XIII. (10.) *The Bonzes* are the minifters of the religion of the Japanefe. Thefe affect great continence, and a wonderful fobriety. They live in community, and have feveral univerfities, where they teach their theology and the myfteries of their fect. Among the Bonzes, there is one named Combadaxi, whom the Japanefe highly revere, and believe him to be immortal. The young women of Japan live alfo in a fort of convents. The name of bonzes is likewife given to fome other priefts among the idolatrous nations of India.

(11.) *The Druids* were the priefts among the ancient Gauls, and they are thought to be the

fame

same with the Eubages, of whom Ammianus Marcellinus speaks, and the Saronides that are mentioned by Diodorus Siculus. They taught a religion to the people, which they had probably learned from the Phoceans. They had an extraordinary veneration for the oak, because that tree bore the mistletoe. For the rest, they applied themselves to the contemplation of the works of nature, and regulated the religious ceremonies, being at once the theologians and philosophers of the ancient Gauls; of whom the *Bards* were the poets, scholars, and musicians.

XIV. (12.) *The Religion of the Peruvians, or the Yncas.* The first king of Peru was, they say, Ynca Manco Capac, and all his successors have been called, from his name, Yncas. The Peruvians make their first kings to be descended from the sun, which they adore as a god. Their other divinities, as the moon, the sister and wife of the sun, which they named Quilla; the star Venus, that they call Chasca; the thunder and lightening, to which they gave the common name of Yllapa; the rainbow, that they named Cuychu; were divinities inferior to the sun. To all these, however, magnificent temples were erected. They sacrificed all sort of animals to the sun, especially sheep, but never men, as the Spaniards have falsely reported of them. They consecrated virgins indeed to the sun, but that was in the manner of devotees, or nuns. These divinities, but especially the sun, had their solemn feasts. The Peruvians,

ruvians, before the Spaniards entered their country, cultivated alſo philoſophy, and eſpecially aſtronomy. It is not wonderful that theſe people, to whom the knowledge of the true God, and of the Chriſtian religion could ſcarce be known, adored the firmament, and eſpecially the ſun, that benign planet, which appears to animate, cheriſh and ſupport all nature. They knew of nothing greater, nothing more worthy of adoration. This worſhip appears, moreover, leſs abſurd than that which the pagans offered to imaginary divinities, or to men whom they had themſelves deified.

XV. Such is nearly the general plan of all the religions that have amuſed the minds of men from the creation of the world to the preſent day. The human mind is conſtantly limited, and its limits are very contracted when it would extend itſelf toward the Supreme Being. We cannot be ſurpriſed therefore, that men of the moſt ſublime genius, and the moſt profound philoſophy, when they have framed new religions, and have aſſumed the important title of leaders of ſects, have laid down falſe ſyſtems, and have frequently united groſs errors and ſuperſtitions with clear, philoſophic truths, and dogmas ſtrictly rational. But while we lament the weakneſs of the human underſtanding, let us remember, that a religion, purely natural and philoſophic, can never ſubſiſt among any nation upon earth; for the bulk of every people cannot, and ought

not,

not, to apply themselves to ratiocination; the state has too much need of their hands, to admit them to apply their heads to abstract speculations. It is therefore absolutely necessary for every founder of a religion, to prescribe a uniform, fixed and immutable standard, as well for the dogmas that the people are to believe, as for the morals they are to practise, and the ceremonies they are to observe in their worship of the Divinity: and this is the more necessary, as the principles of natural religion, if they were alone sufficient to operate the temporal and eternal happiness of mankind, cannot be so fixed, that men of a subtle and philosophic spirit may not, sooner or later, set them in new lights, invent new sects, and throw the whole state into confusion. Let us remember, lastly, that the common people constantly require something marvellous in their religion, and that the marvellous is more difficult to invent than is commonly imagined.

CHAP.

CHAP. IX.

The HISTORY of the CHRISTIAN Church, of Herefies, of the Popes and Reformers.

FROM amidſt the thickeſt darkneſs a light ſhone forth: Jeſus Chriſt, the Saviour of the world, was born at Bethlehem in Judea, on the 25th of December, about the year of the world 4000; in the 23d year of the reign of Auguſtus, and in the 37th of that of Herod. If Chriſt had been nothing more than man, it muſt be confeſſed that he would have been the greateſt of men, the moſt virtuous of the human race, the wiſeſt of philoſophers, and the moſt truly learned of all teachers. His doctrine would not have been leſs divine. He diſcovered to mankind the true and the only principle of all virtue, by ſaying to them, LOVE. But as he is acknowledged by all Chriſtians to be the real Son of God, who came upon the earth to ſave mankind, and offered himſelf as a ſacrifice for the expiation of their ſins, it is not in the power of language fully to expreſs that acknowledgment, that gratitude, veneration, and profound devotion which we owe unto him.

His

His doctrine, his wisdom, his acts, and his miracles, soon distinguished him from all those, who, about the time of his birth, set up for teachers, and assumed the title of King of the Jews, or Messiah; as Theudas, who is mentioned in the Acts of the Apostles, and many others. When he was twelve years old, Jesus was brought by his parents (Mary and Joseph) to the temple of Jerusalem, at the time of the celebration of the feast of Easter: there he seated himself amidst the doctors, who were astonished at his wisdom. From that time he is lost to our sight; he returns to Nazareth, and exercises the profession of a carpenter, with his supposed father, Joseph; earning his bread by the sweat of his brow. When Jesus Christ had attained the age of twenty-six years, John appeared in Judea, declared himself the forerunner of the true Messiah, and baptized Jesus in Jordan, when he was thirty years of age, and was returned from Nazareth in Galilee. The following year Jesus went up to Jerusalem, and there celebrated his first feast of Easter: but hearing that John was imprisoned by Herod the Tetrarch, he left Judea, and returned to Galilee. At the age of 32 years, he went again up to Jerusalem, and there celebrated his second feast of Easter: he then selected his twelve apostles, and afterward retired toward Capernaum: some of his disciples left him, but the apostles remained faithful. The year following, when our Saviour had attained his thirty-third year, he returned to Jerusalem to celebrate his third

third Easter: he then instituted the Holy Supper; was taken into custody by the Jews, was crucified, buried, descended into Hell, rose again, appeared to his disciples, ascended into Heaven, and seated himself on the right-hand of God the Father.

II. They, who would make a regular course in the history of the church, should begin, therefore, by studying the life of Jesus Christ, as it is contained in the four evangelists, Matthew, Mark, Luke, and John, and in the Acts of the Apostles, and no where else. The cotemporary historians make no mention of him: all the traditions that are related are mere fables, without the least authority or appearance of truth; and we must regard these as every man of sense regards the portrait of our Saviour that is said to have been painted by St. Luke, who was a physician; or those relicks of Christ, and of the real cross, of which there are many cart-loads in the world; as it would be very easy to make apparent, did the bounds of this work admit. For the rest, each word that our Saviour pronounced, each act, each miracle that he performed, is a monument of his divine vocation, and which every Christian ought to know and revere.

III. After the death of Christ, his apostles continued to preach his doctrine, and extended it, by degrees, over all the then known world. These twelve apostles were called: 1. Peter, first named

named Simon; 2. James, the son of Zebedee; 3. John, the brother of James; 4. Andrew; 5. Philip; 6. Bartholomew; 7. Matthew; 8. Thomas; 9. James, the son of Alpheus; 10. Jude, or Thades, the brother of James; 11. Simon of Canana; and 12. Matthias, elected by the other apostles in the place of Judas Iscariot, who, after he had betrayed our Saviour, hanged himself in despair. These apostles performed great actions and miracles, which are related by St. Luke in the book called *The Acts*. The apostles chose seven deacons, who were to dispense the alms, and these were, 1. Stephen, a man full of the faith, who was stoned to death; 2. Phillip; 3. Procor; 4. Nicanor; 5. Timon; 6. Parmenas; and 7. Nicholas, a proselyte of Antioch. There were, beside these, seventy-two disciples of Christ, all of whose names are not known to us. By the preaching of the apostles they continually increased, and in process of time the number of proselytes to Christianity, in all countries, was without bounds. Saul, a native of Tarsus in Cilicia, and in that quality a Roman citizen, was a man of distinguished rank, and of great learning. He at first persecuted the Christians, but was soon converted, embraced Christianity, was baptized, and took the name of Paul; he efficaciously assisted the apostles in their labours, and became himself the apostle of the Gentiles. His travels and success are well known. He, and all the other apostles, suffered martyrdom in the first age, except St. John, who died a natural

ral death. Such was the firſt ſtate of the Chriſtian church after its foundation by Jeſus Chriſt. We are likewiſe to examine, in this firſt age, called Apoſtolic, how, when, where, and by whom, the books of the New Teſtament, that is, the Four Evangeliſts, the Acts of the Apoſtles, the Epiſtles or Letters of St. Paul and the other apoſtles, and the Apocalypſe, were written; and by what methods the certainty of their dates, and their authenticity, are eſtabliſhed.

IV. The firſt ages of Chriſtianity were imbrued with blood. We find every where accounts of the troubles, perſecutions and puniſhments which they ſuffered who embraced the Chriſtian doctrine. It ſeems as if the ſovereigns and rulers of the earth had combined to oppreſs this religion, and to exterminate its firſt profeſſors: but Providence was pleaſed to confound the malice and cruelty of man, and even to make the church of Chriſt flouriſh by the blood of the martyrs; to become conſtantly more victorious, and at laſt triumphant, in the fourth century, under the emperor Conſtantine the Great. We learn therefore, in the eccleſiaſtical hiſtory of the firſt three ages, that of the *great perſecutions*, which the emperors and pagan princes made the Chriſtians undergo; and that of the *martyrs*, who ſealed the evangelic faith with their blood; and whoſe names the church has collected in its martyrology.

V. That

V. That we may not confound all those objects which the study of the general history of the Christian church, from its origin to the present time, presents to us, but preserve perspicuity in our ideas of these matters, it seems convenient to make a short analysis of them, by ranging them in the following order: we should therefore make,

VI. (1.) The necessary observations on the first establishment of bishops, and on certain customs of the primitive church. The word bishop comes from the Greek Επισκόπος, and signifies an overseer or inspector: by which is meant a priest, ecclesiastic or sacred prelate, who has the spiritual conduct of a diocese, province or country. He receives his charge by ordination. We find that there were in the primitive church, immediately after the death of Christ, such sort of supervisors or bishops for each particular church, whom St. John in his Apocalypse names, in a figurative style, Angels, as the Angel of Smyrna, the Angel of Laodicea, &c. But these bishops had little resemblance to those of our time: they certainly bore neither mitre nor cross; they did not enjoy the revenues of a prince, nor roll in luxurious pleasures: they lived in the greatest simplicity, instructed, preached, and preserved order among their flocks, without pomp, and without ambition. This part of ecclesiastical history shews also, what were the deacons, deaconesses, and other religious ; the presbytery of bishops, or the college composed of priests and deacons;

deacons; what was the state of the ancient churches and their construction; what is meant by the *agapæ*, or feasts of charity, that were made in the assemblies of the faithful, during the time of the apostles; the *eulogies*, which were small pieces of bread that were sanctified by solemn prayers, to be distributed among the brethren in sign of communion of faith and charity : the *diptychs*, or registers of persons of greatest consequence, who were to be publicly prayed for: the degrees of public penitence* : and lastly, the Eucharist, or holy supper, whose very name and institution prove sufficiently, that it was a solemn supper, which the faithful held among them in order to celebrate the memory of our Saviour; that they made it in their houses, and amidst their families, and not in a church; that it was held at night, and not in the morning, which would have been absurd; that it was never called a sacrament, which is a Latin word, that is not to be found either in the Old or New Testament, but is of modern invention; that it was a repast in which they did not pretend to include any thing mysterious, mystic or miraculous, any more than the Jews did in their paschal lamb, in

* We here discover, in these early days, traces of human invention; for we find not one word of all these matters in the holy scripture. So difficult is it for man to imitate the admirable simplicity of his Divine Master, and to leave his doctrine unaltered, though he has denounced the most terrible imprecations against those who shall add or diminish one word of his gospel.

the

the place of which our Saviour inftituted the holy fupper, by making ufe of almoft the very phrafes, in bleffing the bread and wine, that the father of a Jewifh family made ufe of in bleffing the lamb and the wine of Eafter: in fhort, that it was a refpectable inftitution, but has been ftrangely disfigured.

VII. (2.) *The Hiftory of the Popes.* Though it may appear extraordinary enough, when we form an idea of the prefent popes as heads of the Chriftian church and fecular princes, to find an uninterrupted fucceffion of thefe fovereign pontiffs, from the apoftle St. Peter to Clement XIII. a Venetian; it is, however, convenient and ufeful to follow this feries of the catholic hiftorians, as it produces great order in the hiftory of the church, and leaves no confiderable vacuities to be fupplied. By diftinguifhing, therefore, the eighteen ages of the church, and the reigns of the popes in each century, and by learning the moft confiderable events, with regard to the church, that occurred under each pontificate, we are enabled to acquire a knowledge fufficiently complete of ecclefiaftical hiftory. We can here give their names only, in their proper order.

VIII.
Firft Age.

1. St. Peter the apoftle. 2. St. Linus. 3. St. Cletus, a Roman. 4. St. Clement, a Roman.

Second Age.

5. St. Anaclet, an Athenian. 6. St. Evariſtus. 7. St. Alexander, a Roman. 9. St. Theleſphore, a Grecian. 10. St. Higin, an Athenian. 11. St. Pius of Aquila. 12. St. Anicetus, a Syrian. 13. St. Soter of Fondi. 14. St. Elutherus, a Grecian. 15. St. Victor, an African.

Third Age.

16. St. Zephrinus, 17. St. Calliſtus, 18. St. Urban, 19. St. Pontianus, all Romans. 20. St. Anterus, a Grecian. 21. St. Fabian, 22. St. Cornelius, 23. St. Lucius I. 24. St. Stephen, Romans. 25. St. Sixtus I. 26. St. Denis, both Grecians. 27. St. Felix I. a Roman. 28. St. Eutichian, a Tuſcan. 29. St. Cajus, a Dalmatian. 30. St. Marcellinus, a Roman.

Fourth Age.

31. St. Marcellus, a Roman. 32. St. Euſebius, a Grecian. 33. St. Melchiades, an African. 34. St. Silveſter, 35. St. Mark, 36. St. Julius, 37. St. Liberius, all Romans. 38. St. Damaſcus, a Spaniard. 39. St. Siricus, 40. St. Anaſtaſius I. Romans.

Fifth Age.

41. St. Innocent I. of Albany. 42. St. Zozimus, a Grecian. 43. St. Boniface I. 44. St. Celeſtin I. Romans. 45. St. Sixtus II. 46. St. Leo I. a Tuſcan, ſurnamed the Great. 47. St. Hilary of Sardinia. 48. St. Simplicius of Trivoly.

voly. 49. St. Felix II. a Roman. 50. St. Gelasus, an African. 51. St. Anaftatius, a Roman. 52. St. Simmachus of Sardinia.

Sixth Age.

53. St. Hormifdas, of the Campania of Rome. 54. St. John I. of Tufcany. 55. St. Felix III. of Benevento. 56. St. Boniface II. 57. St. John II. 58. St. Agapitus, all Romans. 59. St. Silverus of Campania. 60. St. Vigil. 61. St. Pelagus I. 62. St. John III. 63. St. Benedict, 64. St. Pelagus II. 65. St. Gregory I. all Romans.

Seventh Age.

66. St. Sabinian of Tufcany. 67. St. Boniface III. a Roman. 68. Boniface IV. of Valeria. 69. Deusdeditus, a Roman. 70. Boniface V. a Neapolitan. 71. Honorius I. of Campania. 72. Severinus, a Roman. 73. John IV of Dalmatia. 74. Theodore of Jerufalem. 75. St. Martin I. of Todi. 76. Eugenius I. 77. Vitalian of Segni. 78. Adeodatus, 79. Domnus, Romans. 80. St. Agathen, 81. St. Leo II. both of Sicilia. 82. St. Benedict II. a Roman. 83. St. John V. a Syrian. 84. St. Conon of Tarfus. 85. St. Sergius, a Syrian.

Eighth Age.

86. John VI. 87. John VII. both Grecians. 88. Sifinnius, 89. Conftantine, both Syrians. 90. Gregory II. a Roman. 91. Gregory III. a Syrian.

rian. 92. St. Zachary, a Grecian. 93. St. Stephen II. 94. St. Stephen III. 95. Paul I. a Roman. 96. Stephen IV. a Sicilian. 97. Adrian I. 98. Leo III. Romans.

Ninth Age.

99. Stephen V. 100. Pascal I. 101. Eugenius II. 102. Valentianus. 103. Gregory IV. 104. Sergius II. 105. Leo IV.* 106. Benedict III. 107. Nicholas I. called the Great, all Romans. 108. Adrian II. 109. John VIII. a Roman. 110. Martin II. a Tuscan. 111. Adrian III. 112. Stephen VI. 113. Formosus. 114. Boniface VI. 115. Stephen VII.

Tenth Age.

116. John IX. of Trivoli. 117. Benedict IV. a Roman. 118. Leo V. of Ardea. 119. Christopher, a Roman. 120. Sergius III. of Tusculum. 121. Anastasius III. a Roman. 122. Lando Sabinus. 123. John X. 124. Leo VI. a Roman. 125. Stephen VIII. 126. John XI. of Tusculum. 127. Leo VII. a Roman. 128. Stephen IX. a German. 129. Martin III. a Roman. 130. Agapitus II. a Roman. 131. John XII. of Tuscany. 132. Benedict V. 133. John XIII. 134. Domnus II. 135. Benedict VI. all Romans. 136. Benedict VII. 137. John XIV. of Pavia. 138. John XV. a Roman.

* Between Leo IV. and Benedict III. some place pope Joan, under the title of John VII.

139. Gre-

139. Gregory V. a German. 140. Silvester II, a monk of Auvergne.

Eleventh Age.

141. John XVI. 142. John XVII. 143. Sergius III. Romans. 144. Benedict VIII. 145. John XVIII. 146. Benedict IX. Tuscans. 147. Gregory VI. 148. Clement II. 149. Damasus II. 150. Leo. IX. 151. Victor II. Germans. 152. Stephen X. of Lorrain. 153. Nicholas II. a Savoyard. 154. Alexander II. of Lucca. 155. St. Gregory of Soana. 156. Victor III. of Benevento. 157. Urban II. a Frenchman. 158. Pascal II. a Tuscan.

Twelfth Age.

159. Gelasus of Gaita. 160. Calistus II. of Bourgogne. 161. Honorius II. of Boulogne. 162. Innocent II, a Roman. 163. Celestin II. a Tuscan. 164. Lucius II. of Boulogne. 165. Eugenius III. of Pisa. 166. Anastasius IV. 167. Adrian IV, an Englishman. 168. Alexander III. of Sienna. 169. Lucius III. of Lucca. 170. Urban III. of Milan. 171. Gregory VIII. of Benevento. 172. Clement III. a Roman. 173. Celestin III. a Roman. 174. Innocent III. of Anagnia.

Thirteenth Age.

175. Honorius III. a Roman. 176. Gregory IX. of Anagnia. 177. Celestin IV. of Milan. 178. Innocent IV. of Genoa. 179. Alexander IV,

IV. of Anagnia. 180. Urban IV. of Troja. 181. Clement IV. of St. Giles's. 182. Gregory X. of Plaifance. 183. Innocent V. of Lyons. 184. Adrian V. count of Lavagne. 185. John XIX. of Frefcati. 186. Nicholas III. of Rome. 187. Martin IV. of Brey. 188. Honorius IV. of Rome. 189. Nicholas IV. of Afcoli. 190. Celeftin V. of Ifernia. 191. Boniface VIII. of Anagnia.

Fourteenth Age.

192. Benedict X. of Trevifa. 193. Clement V. of Bazas. 194. John XX. commonly called John XXII. of Cahors. 195. Benedict XI. of Foix. 196. Clement VI. of Limofin. 197. Innocent VI. of Limofin. 198. Urban VI. of Manda. 199. Gregory XI. of Limofin. 200. Urban VI. a Neapolitan. 201. Boniface IX.

Fifteenth Age.

202. Innocent VII. of Sulmona. 203. Gregory XII. a Venetian. 204. Alexander V. of Candia. 205. John XXI. commonly called the XXIII. a Neapolitan. 206. Martin V. a Roman. 207. Eugenius IV. a Venetian. 208. Nicholas V. of Lucca. 209. Califtus III. a Spaniard. 210. Pius II. of Sienna. 211. Paul II. a Venetian. 212. Sixtus IV. of Savona. 213. Innocent VIII. of Genoa. 214. Alexander VI. a Spaniard.

Sixteenth

Sixteenth Age.

215. Pius III. of Sienna. 216. Julius II. of Savona. 217. Leo X. a Florentine. 218. Adrian VI. of Utrecht. 219. Clement VII. a Florentine. 220. Paul III. a Roman. 221. Julius III. a Tuscan. 222. Marcellus II. 223. Paul IV. a Neapolitan. 224. Pius IV. of Milan. 225. Pius V. of Alexandria. 226. Gregory XIII. of Boulogne. 227. Sixtus V. of Ancona. 228. Urban VII. 229. Gregory XIV. of Milan. 230. Innocent IX. of Boulogna. 231. Clement VIII. of Florence.

Seventeenth Age.

232. Leo XI. of Medicis, a Florentin. 233. Paul V. a Borgesian. 234. Gregory XV. 235. Urban VIII. a Florentine. 236. Innocent X. a Roman. 237. Alexander VII. of Genoa. 238. Clement IX. of Pistonia. 239. Clement X. a Roman. 240. Innocent XI. of Milan. 241. Alexander VIII. of Rome. 242. Innocent XII. a Roman.

Eighteenth Age.

243. Clement XI. of the dutchy of Urbano. 244. Innocent XIII. a Roman. 245. Benedict XII. or XIII. by the reason of the antipope Benedict. 246. Clement XII. a Florentine. 247. Benidict XIV. and 248. Clement XIII. a Venetian.

IX. How

IX. How happy, how glorious would it have been for Chriftianity if all thefe heads of the vifible church, all thefe vicars of Chrift, had been animated with the fpirit of their Divine mafter; if they had been fagacious, learned, wife and virtuous; if they had all refembled Benedict XIV. and Clement XIII. But fuch was not the will of Providence, for the tiara has been frequently born by the moft criminal heads. It is not for us, however, to fcrutinize the counfels of the Supreme Being, nor to be diffatisfied with thofe inftruments of which he has thought proper to make ufe, in executing his eternal decrees.

X. (3.) *The Hiftory of the Schifms* that have arofe in the Chriftian church, and efpecially that grand divifion by which it was divided into the Greek and Latin churches. This fchifm began about the year of Chrift 854. under the emperor Michael of Conftantinople. Its origin and progrefs are to be found in all the hiftorians; but to form a juft judgment it is neceffary to read the authors of both parties. The empire of the Eaft has followed from that time the dogmas and rites of the Greek church, and the empire of the Weft the dogmas and rites of the Latin. The empire of the Eaft being now in the hands of the Mahometans, it is only the Greeks in Europe, in Afia Minor, and the iflands; the Syrians, the Georgians, and the Ruffians, who form the Greek church, under the patriarchs of Conftantinople, Alexandria, Antioch, Jerufalem and Ruffia.

Ruffia. The patriarch of Conftantinople bears the title of *Panagiotita fou*, or his *all-holinefs*. There are in this church archimandrins or abbies, archbifhops, bifhops, fuffragans, bapas or curates, and the religious named caloyers, who wear a black habit, nearly the fame as that of the Benedictines. Ecclefiaftic hiftory informs us what are the particular circumftances that have attended the ancient church; the fucceffion of its patriarchs; the councils it has held; and what are its dogmas, its rites and ceremonies.

XI. *The Hiftory of the Councils*, during the eighteen centuries of the univerfal Chriftian church. Thefe councils have been either œcumenical, in which all Chriftianity is interrefted; or national, or provincial, or diocefian; and the conciliabules, held by fchifmatic ecclefiaftics. They call the firft council that affembly of the apoftles held in Jerufalem, where Jofeph, Barfabas and Matthias, were propofed to fill the place of Judas the traitor, when the lot fell on Matthias. There have been fince that time many of thefe forts of affemblies of bifhops and principal ecclefiaftics, which may be compared to provincial councils, but have never borne that title. The firft general council was held at Nice, a city of Bithynia, in Afia Minor, in the year 325. The œcumenical councils which have fucceeded that, are

2. That of Conftantinople, held in the year 381.

3. That

3. That of Ephesus, in the year 431.
4. That of Calcedonia, 451.
5. The second of Constantinople, in 553.
6. The third of the same city, 680.
7. The second of Nice, in the year 787.
8. The fourth of Constantinople, in 869.
9. That of the Lateran, held in 1123.
10. The second of the Lateran, in the year 1139.
11. The third of the same place, in 1179.
12. The fourth of the same place, in 1215.
13. That of Lyons, held in 1245.
14. The second of Lyons, in 1274.
15. That of Vienna, in 1311.
16. That of Constance, in 1414.
17. That of Basil, in 1431.
18. That of Florence, in 1439.
19. The fifth of the Lateran, in 1512, and lastly,
20. That of Trent, held in the year 1545.

The decisions of these councils are named decrees or canons, and are regarded as infallible, because they are supposed to have been immediately dictated by the Holy Spirit. The Holy Spirit, however, has not been accustomed to revoke and contradict its decrees, as these councils have publickly done. When the council of Trent was sitting, there were in the world certain wicked wits, who said that the Holy Spirit arrived at Trent every day in the cloak-bag of the postillion who came from Versailles. By reading with attention the history, the debates, and decrees of all these councils, we may see the origin

gin and date of each article of faith, and each dogma, contained in the theory or dogmatic, and in the catechisms of modern Christians; as they have deduced them from the principles of the gospel.

XII. *The History of the Heresies.* Any doctrine that is contrary to the decisions of the catholic church is called a heresy: an heresiarch, therefore, is one who invents and maintains such doctrine, and from whom it takes its name; and a heretic is he who embraces and follows that doctrine. According to this definition we cannot say that there have been any heretics, properly so called, since the time of the apostles, because the general system of the catholic religion, as it is at this day, has been formed by the successive decisions of the councils only: for otherwise, a man who had advanced, in the second or third century, a point of doctrine that was not established by any council till the tenth century, would have been a heretic. This is so clear, that it is not likely any one will be hardy enough to deny it. But if we agree to call those heretics who have advanced and maintained doctrines contrary to those received and taught by the Christian church at that very time, there have been certainly heretics without number in all ages of the church, from Simon the Magician and Dosithea, who lived in the time of the apostles, down to count Sintzendorff, leader of the Herrenhutters. Ecclesiastic history informs us, from age to age,

what

what were their names, their doctrines, the progress that they made, and the obstacles they encountered. It makes particular mention of one named Manes, who lived in the third century, about the year 277, and who was the founder of a sect called Manicheans: of one Arius, who appeared in the fourth century, and became the chief of the formidable sect of Arians: of one Pelagius, who established, at the beginning of the fifth century, the celebrated sect of Pelagians: of one Nestorius, who, about the year 430, founded the sect of Nestorians: of one Lelius Socinus, who formed, toward the middle of the sixteenth century, the sect of Socinians; and of many other heresiarchs, who have made themselves very famous in the world.

XIII. But it appears to be unjust to give the odious appellation of heresiarch, or heretic, to Martin Luther, or John Calvin; who, far from attempting to introduce any new dogmas into the church, have not, in any manner, attacked the fundamental principles of the Christian religion; but have applied themselves solely to the re-establishment of the pure and simple doctrine of Jesus Christ, and to the purging of the catholic religion from divers points of faith and practice, which time, the troubles of the church, its leaders and councils, had introduced, and which had rendered the doctrine of our Saviour quite different from that simplicity and humility by which it was originally characterized. Their only intention was to reform abuses, and not to introduce new systems. Perhaps they wanted either discernment or courage, or proper support, to retrench more of those dazzling superstitions. Be that as it may, ecclesiastic history instructs us in,

(6.) *The History of the Reformation*, in its full extent, as well as the most remarkable events that have occurred in the two religions (the Calvinist and Lutheran) from the time that some of the principal nations of Europe have embraced them.

XIV. This history likewise informs us,

(7.) *Of the different Sects which at this Day divide the three principal Christian Communities*, who, though they follow in general the fundamental dogmas of their communion, and the rites of their church, yet differ from it in some essential articles. Such are the Molinists and Jansenists among the Catholics; the Moravian brethren, or Herrenhutters, among the Lutherans; and the Armenians, Gomarists, Coccigans, &c. among the Calvinists. We here see also the origin and history of the Mennonists and the Quakers, and, in short, of all the sects which now subsist in the Christian world.

XV. (8.) *The History of the Martyrs.* Mankind have been, in all ages, so senseless and inhuman as to persecute their brethren for seeking a different way by which they might arrive at eternal felicity, and have even carried their barbarity so

far

far as to cause them, for that reason only, to expire in tortures: an absurdity as great, a practice as enormously inhuman and wicked, as it would be to put them to the torture for going to Paris or Rome by a road different from that which is taken by the post. The first Christians, in particular, endured inexpressible, inconceivable persecutions and torments. They whose blood has been spilt in the cause of religion are called Martyrs, and their names, as well as the history of their lives and deaths, are recorded in those immortal books called Martyrologies. There are some of these that contain merely a list of their names, and of the place and day of martyrdom of each saint. Baronius gives to pope Clement I. the glory of having introduced the custom of collecting the acts of the martyrs. The martyrology of Eusebius of Cæsarea, which is attributed to St. Jerom, is the most ancient that is known to us. That of Beda was wrote in 730. The ninth century was very fruitful of works of this sort. There is also the small martyrology that was sent by the pope to Aquila; those of Florus, Wandelbert, Raban, Notker, Adon, Usuard, Nevelon, Ditmar, &c. The martyrologies were preceded by the calendars.

XVI. (9.) *The History of the Religious Orders.* By which is meant such societies of religious people as monks and nuns, who live under the direction of a chief, observe the same regulations, and wear the same habit. These religious orders may be reduced to five claſſes, monks, canons, knights, mendicants, and regular clerks. Many of the fathers of the church regard St. John the Baptiſt as the founder of a monaſtic life, and St Jerom calls him, on that account, *monachorum princeps*. But nothing is more ridiculous than ſuch an opinion. What reſemblance is there between St. John and a monk? Could St. John ever think of prohibiting that which God and religion, poſitive and natural, permit; that is, the allowing of churchmen to marry, and provide inhabitants for the world, and ſubjects for the ſtate? Be this however as it may, we find in the hiſtory of the church (eſpecially in thoſe that are wrote by catholic authors) a ſeries of all the religious orders that have been founded in Chriſtianity during the eighteen centuries that it has ſubſiſted, with the regulations that each of theſe orders have adopted and followed. Father Helyot, a penitent of the third order of St. Francis, has formed a hiſtory of the monaſtic, religious and military orders, and of all the ſocieties of each ſex: and there is, at the beginning of his firſt volume, a catalogue of ſuch books as treat of theſe orders.

XVII. (10.) *The Series of the principal Authors of Sacred Hiſtory.* At the head of this laſt diviſion are naturally placed,

1. The ſacred authors of the New Teſtament. Our Saviour has left us no part of his divine

divine doctrine in writing. The whole of it was collected and digested by the four evangelists. St. Luke wrote the Acts of the Apostles, and St. John the Apocalypse. The rest of the New Testament consists of epistles or letters, which St. Paul, St. James, and St. Jude wrote, after the death of Christ, to some churches of the faithful, or to some of their relations.

2. The fathers of the church. By this title is properly meant those ecclesiastical writers who have preserved what is called the tradition of the church. Their writings are held in high veneration, and have an extraordinary authority in the catholic church, and are in much esteem among the other communions. The catalogue of these is to be found in most ecclesiastic histories, but is too numerous to be inserted here.

3. The other catholic authors, who have wrote since the beginning of the thirteenth century, down to the present day, on matters of importance to religion, and who are called doctors.

4. The principal Lutheran authors, from Martin Luther, Phil. Melanchton, &c. to the present time.

5. The

5. The principal writers among the Calvinists, from John Calvin, Zuinglius, Oecolampadus, &c. down to our own day.

6. The Socinian authors, who are also called Polonian brethren, whose works have been collected; as those of Socinus, Crellius, Walzogen, &c.

7. The Jansenist and Molinist writers, &c. among the catholics: and lastly,

8. The writers among the various modern sects, as Quakers, Mennonists, Herrenhutters, &c.

He, who shall study ecclesiastical history according to the plan we have here laid down, will acquire, we apprehend, a complete knowledge of it, and at the same time range in his memory all its various matters, in a proper order.

CHAP. X.

ANTIQUITIES.

WE should not confound in our ideas the different terms of Antiquities and Antiques. By antiquities are meant all testimonies or authentic accounts, that have come down to us, of ancient nations; and by antiques, those precious works in painting, architecture, sculpture, and graving, that were made from the time of Alexander the Great, to that of the emperor Phocas, and the devastations of the barbarians; that time has spared and has committed to our care, and which are the ornaments of our cabinets and galleries, and sometimes of the gardens of princes. Antiques therefore make only a part of antiquities, and the latter form a very extensive science, including "an historical knowledge of the edifices, magistrates, offices, habiliments, manners, customs, ceremonies, worship, and other objects worthy of curiosity, of all the principal ancient nations of the earth."

II. This science, therefore, is not a matter of mere curiosity, but is indispensable to the theologian;

logian; who ought to be thoroughly acquainted with the antiquities of the Jews, to enable him properly to explain numberless passages in the Old and New Testament: to the lawyer; who, without the knowledge of the antiquities of Greece and Rome, can never well understand, and properly apply, the greatest part of the Roman laws: to the physician and the philosopher, that they may have a complete knowledge of the history and principles of the physic and philosophy of the ancients: to the critic, that he may be able to understand and interpret ancient authors: to the orator and poet; who will be thereby enabled to ornament their writings with numberless images, allusions, comparisons, &c. all which gave Masenius occasion to say: *Quicunque ad aliquam inter Romanos eloquentiæ facultatem adspirat, hanc veterem Romanæ urbis historiam, originem, mores, instituta hujus gentis, disciplinam in toga sagoque usitatam, tenere necesse est. Neque enim citra hanc cognitionem priscos Romanæ eloquentiæ assertores, Ciceronem, Livium, Plinium, Terentium, aliosque, satis quisquam vel legendo assequatur, vel imitetur scribendo. Palæstr. Styli Rom. L. III. c. 18.*

III. Antiquities are divided into sacred and profane, into public and private, universal and particular, &c. It is true that the antiquaries (especially such as are infected with a spirit of pedantism, and the number of these is great) frequently carry their inquiries too far, and employ them-

themselves in laborious researches after learned trifles: but the abuse of a science ought never to make us neglect the applying it to rational and useful purposes.

IV. Many antiquaries also restrain their learned labours to the ecclaircissement of the antiquities of Greece and Rome: but this field is far too confined, and by no means contains the whole of this science, seeing it properly includes the antiquities of the Jews, Egyptians, Persians, Phenicians, Carthaginians, Hetruscans, Germans, and, in general, all those principal nations whom we have mentioned in the 5th chapter of ancient history; so far as any accounts of them are come down to us.

V. It will be easily conceived, that it is not possible for us to enter here into the detail of all these matters: it is our business, however, to inform our readers of what they ought to inquire after in the study of the antiquities of each people, as far as the monuments or memoirs that are yet remaining can furnish any lights; and this is what remains to be done to complete this chapter.

VI. The science of antiquities includes therefore,
1. The origin of a people, and of the name they bear.
2. The

2. The local situation of the country they inhabited.
3. The extent and bounds of their country.
4. The climate and its properties.
5. The genius and spirit of the people.
6. Their manners.
7. The progress they have made in arts and sciences, in commerce, navigation, &c.
8. Their military capacity; their valour, discipline, knowledge in fortification, &c.
9. The geographic description of the country; its mountains, forests, rivers, lakes, &c.
10. The natural history of the country; its animals, plants, minerals, and other productions.
11. The account of its principal cities, and especially its capital.
12. Its bridges, gates, highways, and most considerable edifices.
13. Its public places.
14. Its aqueducts, cisterns, fountains, &c.
15. The palaces of its kings, princes, or senate.
16. All its other public buildings, as arsenals, tribunals of justice, public halls, &c.

VII. And also,
17. The public libraries.
18. Public baths.
19. Harbours and keys.

20. The-

20. Theatres, amphitheatres, circuses, places for public combats, &c.
21. Subterraneous passages for water, as common sewers, &c.
22. Public magazines and granaries.
23. Public schools.
24. The fields where the soldiery were exercised.
25. The public mills.
26. The burses, or places where the merchants assembled.
27. The houses of private persons, as well in town as country.
28. Their carriages, cars, litters, &c. ecuries, &c.

VIII. Embellishments and ornaments in architecture and statuary, as
29. Triumphal arches.
30. Columns.
31. Obelisks.
32. Colosses.
33. Equestrian and pedestrian statues, groups, &c.
34. Bass-relieves, &c. To all which should be added inquiries into the mechanics of the ancients, or the machines of which they made use in their immense works, and the advancement they had made in this art.

IX. Sacred

IX. Sacred antiquities, comprehending,
35. Their temples, chapels, sacred groves, &c.
36. The gods of each nation, their demigods, &c.
37. The general and particular worship of each people.
38. Their idols, oracles, &c.
39. Their priests, sacrificers, augurs, flamens, and other persons of both sexes employed in their sacred offices.
40. Their solemn feasts, and particularly those instituted in honour of each divinity.
41. The habiliments and ornaments of the priests and ecclesiastics.
42. The vases, censers, altars, and utensils that they employed in the sacred service.
43. Their sacrifices and victims.
44. Mysteries.
45. Sacred books.
46. Lares or domestic gods.
47. Processions. And lastly,
48. The principal dogmas of the religion, and the precepts of morality of each people.

X. In profane antiquities, there are likewise to be inquired after,
49. The public shews that were exhibited by the ancient nations in general.
50. Their tragedies, comedies, mimes, pantomimes, &c.

51. Their

51. Their games, as the olympic and capitolian games: their fairs, &c.
52. The combats of gladiators, wrestlers, wild beasts, &c.
53. The races of men and horses.
54. The music of the ancients, and the instruments that were in use among each people.

This division likewise includes their triumphs, and the several crowns and diadems with which they ornamented the heads of their emperors, kings, conquerors, priests, priestesses, poets, and other illustrious personages, &c.

XI. They next pass to the examination of political subjects, as
55. The form of government.
56. The division of a people into tribes.
57. The chiefs of each people, and their authority.
58. The heads of their tribes.
59. Their magistrates.
60. Their manner of rendering justice, and the method of process in their laws.
61. Their criminal justice.
62. The corporeal punishments, and other political pains, penalties, and ignominies which they inflicted.
63. The various classes of the inhabitants; as, among the Romans, the patricians, knights, plebeians, senators, the people in a body, the

the nobles, ignobles, the *ingenui*, the freed-men and the *libertini*.

64. Their flaves; the nature of flavery, fervitude, and of the peculium or property of flaves.
65. Their ambaffadors and their privileges.
66. Their military officers of all ranks; the nature of their troops, their duty, and of the art of war among them.
67. The civil laws of each people.
68. Their criminal laws.
69. The public conftitution of each nation.
70. The affemblies of the people, and their deliberations on the affairs of ftate.
71. The nature of the finances of the ancient nations, and of their contributions.
72. The induftry of the people, their manufactures and commerce.
73. Their mines, and the manner of working them.
74. Their agriculture and rural economy.
75. Their weights and meafures.
76. Their current coins, and their value.
77. Their medals, and their ufe.
78. The folemn forms which they obferved, as well in their public acts, as in their contracts, wills, and other private affairs.

XII. In the laft place, they examine into certain ufages and cuftoms obferved by ancient nations, in private life, as

79. Their

79. Their marriages.
80. Their burials, sepulchres, funeral urns, &c.
81. The ordinary dress of the inhabitants of both sexes; their manner of cloathing the head, body, and feet; and the ornaments of their dress, &c.
82. Their different kinds of foods, and methods of preparing them.
83. Their manner of sitting at table.
84. Their ordinary drink, and strong liquors.
85. Their beds, dormitories, furniture and utensils.
86. Their chests and cabinets.
87. The proper names of the ancients, and especially those of the Romans, who had several, as Marcus Tullius Cicero: and an infinity of other like matters, as,
88. The education they gave their children, &c.

XIII. If to all these general subjects we add the particular study of antiques, of the statues, bail-relieves, and the precious relicks of architecture, painting, camaycus, medals, &c. it is easy to conceive that antiquities form a science very extensive and very complicate, and which is alone sufficient to employ the whole life of a man who is a laborious student: and though a strong memory be the principal faculty that is required, yet great sagacity and attention are necessary in comparing the several objects, in drawing judicious inferences,

inferences, and in forming from thence an ingenious and rational fyftem. It is manifeft, moreover, that the ftudy of antiquities muft be vaftly extenfive, when we confider that all the articles we have enumerated for one people, muft be extended to all the nations of antiquity; and that we muft know them, as if, in a manner, we had lived among them. But this is a knowledge that it would have been impoffible for any one man whatever to have attained, if our predeceffors had not prepared the way for us; if they had not left us fuch ineftimable works as thofe of Gronovius, Grævius, Montfaucon, count Caylus, Winckelmann, the Hebraic antiquities of D. Iken of Bremen, the Grecian antiquities of Brunings, the Roman antiquities of Nieupoort, and efpecially that work which is intitled Bibliographia Antiquaria Joh. Alberti Fabricii, profeffor at Hamburg.

XIV. Nor muft we here forget that very valuable work, with which Mr. Robert Wood, an Englifhman, has lately enriched this fcience, and which is fo well known, and fo juftly efteemed by all true connoiffeurs, under the title of the *Ruins of Palmyra*, and thofe of *Balbeck*. It is by this work that we are fully convinced of the grandeur and magnificence, the tafte and elegance of the buildings of the ancients. We here fee that the invention of thefe matters is not all owing to the Greeks, but that there were other nations who ferved them as models. For

though

though many of the edifices of Palmyra are to be attributed to the emperor Aurelian, and to Odenatus and his wife Zenobia, who reigned there about the year 264, yet there are found, at the same place, ruins of buildings, that appear to be of far greater antiquity, and that are not less beautiful. The ancient Persepolis is sufficient to prove this assertion. When we seriously reflect on all these matters, and especially if we attempt to acquire any knowledge of this science, we shall soon be convinced that it but ill becomes a petit-maitre to laugh at a learned antiquary.

XV. The knowledge of these monuments of the ancients, the works of sculpture, statuary, graving, painting, &c. which they call *antiques*, requires a strict attention, with regard to the matter itself on which the art has been exercised; as the wax, clay, wood, ivory, stones of every kind, marble, flint, bronze, and every sort of metal. We should begin by learning on what matter each ancient nation principally worked, and in which of the fine arts they excelled. For the matter itself, as the different sorts of marble, compositions of metals, and the species of precious stones, serve frequently to characterize the true antique, and to discover the counterfeit. The connoisseurs pretend also to know, by certain distinct characters in the design and execution of a work of art, the age and nation where it was made. They find, moreover,

over, in the invention and execution, a degree of excellence, which modern artists are not able to imitate. Now, though we ought to allow, in general, the great merit of the ancients in the polite arts, we should not, however, suffer our admiration to lead us into a blind superstition. There are pieces of antiquity of every sort, which have come down to us, that are perfectly excellent, and there are others so wretched, that the meanest among modern artists would not acknowledge them. The mixture of the good and bad has taken place in all subjects, at all times, and in all nations. The misfortune is, that most of our great antiquaries have been so little skilled in designing, as scarcely to know how to draw a circle with a pair of compasses. It is prejudice therefore, which frequently directs them to give the palm to the ancients, rather than a judgement directed by a knowledge of the art. That character of expression, which they find so marvellous in the works of antiquity, is often nothing more than a mere chimera. They pretend that the artists of our days constantly exaggerate their expressions; that a modern Bacchus has the appearance of a man distracted with intoxication; and that a Mercury seems to be animated with the spirit of a fury, and so of the rest. But let them not decide too hastily. Almost all the antique figures are totally void of all spirit of expression; we are forced to guess at their characters. Every artificial expression requires, moreover,

to

to be somewhat exaggerated. A statue or portrait is an inanimate, a dead figure, and must therefore have a very different effect from one, which, being endowed with life, has the muscles constantly in play, and where the continual change of the features, the motion of the eyes, and the looks, more or less lively, easily and clearly express the passions and sentiments. Whereas in a figure, that is the produce of art, the delicate touches, that should express the passions, are lost to the eyes of the spectators: they must therefore be struck by strong, bold characters, which can affect them at the first glance of the eye. A very moderate artist is sensible, at the same time, that he is not to give his figures extravagant expressions, nor to place them in distorted attitudes.

XVI. We will finish this chapter with one material observation. All the sciences, by which we can acquire any knowledge of antiquity, as, 1, That which we have here explained; 2, that of medals and coins; 3. the diplomatic, and the explication of inscriptions, or what is called *Epigrammatographica*, or *res lapidaria*; and 4, The knowledge of books, are comprised under the common collective title of *Literature*. But by a caprice of the literati, they have included, under that denomination, the philosophic sciences and history: though for so doing, there can be no good reason whatever. Why should we perplex the ideas of those who are desirous of obtaining a knowledge of these matters, by confounding the sciences? Ought we not much rather to endeavour carefully to mark their distinct limits? But perhaps their intention is to comprehend, under the denomination of literature, the whole of Universal Erudition; and if that be the case, we are not desirous of disputing with any one about words.

CHAP. XI.

Of Medals and Coins.

I. WE shall begin with coins, because they are most ancient, and of most universal use; money was current a long time before they had invented the method of preserving the memory of illustrious persons, by those little monuments of metal, which imitate coins, and are easily dispersed among mankind, and which are called Medals. The number that has been made of these medals is, beside, vastly inferior to that of monies; and the coins of the ancients are, moreover, become our medals. The

The learned comprehend these two objects, which form an important part of literature, under the denomination of Res Nummaria, or Numismatica.

II. It is certain, that in the most ancient times, all commerce was carried on by barter. There was always a necessity, however, for a sort of common measure, by which they estimated the value of commodities. The first inhabitants of the earth were almost all shepherds and husbandmen: they therefore made that common measure to consist of a certain portion of their flocks, which was considered *tanquam pretium eminens*: and any commodity was said to be worth so many oxen, sheep, &c. as is confirmed by Gellius, *Noctes Atticæ*, l. xi. c. 1. In process of time, they found it more expedient to express the value of most commodities, by bits of leather, which by their marks showed the number of beasts they were worth. This was the first money, and the origin of all coins. History says positively, that Numa Pompilius caused money to be made of wood and leather: and from hence came the Latin word *pecunia*. Cassiodorus says likewise in express terms: *Pecunia enim a pecudis tergo nominata, Gallis auctoribus, sine aliquo adhuc signo ad metalla translata est.* He treats also *de assibus scorteis* in the tenth book.

III. Metals being found the most incorruptible of all substances, they afterwards made use of

bits

bits of rough copper in the room of leather, which they called *es rude*, and reckoned by their weight: thefe were after marked according to their weight, and laftly, with images. And we ftill fee, on the moft ancient coins, the figures of animals, and efpecially of oxen and fwine. Numa, toward the end of his reign, began to caft money, and it was from him that came the word Nummus. They formed pieces of money of different weights, and marked on each, as we have juft faid, its weight, or its intrinfic value. It is time that perfects all inventions, and it was time that taught the ancient nations (as it may one day teach the modern Swedes) that the precious metals were more commodious in the commerce of life, and that a lefs weight might exprefs, and be equal to, a greater value; and from difcovering this, they came to form money of filver and gold.

IV. But, in the daily ufe of thefe pieces, it would be impoffible always to weigh them, and much fraud might arife by depending on their marks. To obviate this inconvenience, the fovereigns of each country took on them the exclufive office of making money; and that the public might be certain the weight was juftly marked, they ftamped them on one fide with their image, and on the other with their arms or cypher: which practice has continued to the prefent day: and it is manifeft, that the credit and glory of a prince is concerned in having

having the coin, which bears his image, contain the true value, both with regard to the present age, and to posterity.

V. Mankind have also contrived to preserve the memory of great events, and of illustrious personages, by coins which they call *medals:* a term that is manifestly derived from the word metal. These precious monuments of antiquity do not, therefore, serve merely to engage the curiosity of the scholar and the connoisseur, but are of use also in elucidating history; in fixing the chronology, and in throwing clear lights on ancient events: and as the current coins of antiquity cannot pass among us, on account of the small number that is remaining of them, and of the difference in value of gold and silver, these coins are now become the most precious medals.

VI. The greatest part of antique coins and medals, especially the Greek and Roman, are so finely struck, the design and graving so perfect, the invention simple and sublime, and the taste so exquisite, that independent of their utility in history, we cannot sufficiently admire their intrinsic merit, and must constantly regard them as incontestable proofs of the perfection of the arts in those distant ages. It is not therefore wonderful, that so many persons of discernment, taste, and learning, have employed themselves in forming collections of the coins and medals of the ancients;
and

and that so many learned men have wrote curious and inftructive treatifes concerning them; and laftly, that the knowledge of thefe precious monuments is become a very extenfive branch of fcience, under the title of Numifmatographia; and which we fhall now endeavour briefly to explain.

VII. Medals may be divided into different claffes,

(1.) According to the time when they were ftruck: and in this refpect they are either,

1. Antiques; which are thofe that were made from the moft ancient times of which we have any account, down to the fixth or feventh century of the Chriftian era.

2. Thofe of the middle age; which is from the feventh century, or the death of Phocas and Heraclius, in 641, when Italy became a prey to the Barbarians; where thofe fine medals that are called *Imperials* end, and where begin thofe of the lower empire, and of the Grecian emperors, down to the taking of Conftantinople. The *Gothics* continue the feries from the Imperials. They are fo called, becaufe they were made in the time of the Goths, during the decline of the two empires; and they refemble the ignorance of their age. The connoiffeurs pay but little regard to thefe: they are, however, of great importance in hiftory, in afcertaining the true chronology of events. Thefe come quite down to the fifteenth century.

3. The

3. The modern; which are thofe that have been ftruck in Europe, from the time that the Goths were exterminated, and the art of engraving began again to flourifh. The firft of thefe is that of John Hufs, a famous heretic, which was ftruck in the year 1415. This art has rofe with great luftre from its afhes: there are now many excellent medallifts, and we have feen pieces executed by the celebrated Hedlinger, a Swede, which, prejudice apart, are nothing inferior to the moft finifhed that Greece and Rome have left us.

VIII. (2.) According to the nature and quality of the metal: and in this refpect they are either of

1. Gold; whofe feries is the leaft numerous, and fcarce exceeds 1000 or 1200 in the imperials.

2. Silver; the feries of which may amount to 3000, in the imperials alone.

3. Brafs; which are of three different fizes, that are called the great, the middle, and fmall, and of which the feries amounts to 6 or 7000, if not more, in the imperials. It is not, however, either the metal or the magnitude that renders medals valuable, but the rarity of the head, the reverfe, or the legend. A medal may be common in gold that is very fcarce in brafs; or very rare in filver, that is common in gold or brafs. A head may be common that has a very uncommon reverfe, and the contrary. There

are

are also medals that are only scarce in some series, and very common in others, as in gold, silver, the great, middle, or small brass.

IX. (3.) According to their essential qualities, and the use to which they have been applied: and in this respect they are,

1. Coins, that have anciently served in the commerce of life, but which time has rendered medals.

2. Real medals, struck in the form of coins, either in gold, silver or brass, to preserve to posterity the image of illustrious persons, or the memory of some important action.

3. Medallions; which are properly nothing more than medals uncommonly large; and which have been presented by princes to those whom they have honoured with their esteem: or to serve as public monuments. The Romans named them Missilia. There is no series to be formed of these, even if the different magnitudes and metals be united: and there are not above four or five hundred of them to be found in the richest cabinets.

X. (4.) According to the nation by whom they have been made: and in this case they are,

1. Hebraic. The common opinion is, that there are no Hebrew medals, and that the Jews learned the knowledge of them from the Romans, when they invented the art of clipping them.

them. But, as we have said in the first section, the coins of the ancients are become our medals, and especially the Hebraic, which are also called Samaritan, because their legend is usually in the Samaritan language, and there is reason to believe that there was a mint in that city. There are twenty passages in the Bible which prove that the Jews knew the use of money in the time of Solomon. In the cabinets of the curious there are to be found shekels of copper or silver, and we are assured that there is a gold Hebraic medal in the cabinet of the king of Denmark: but this is the only one that is known. Father Soucier has wrote a dissertation on Hebraic or Samaritan medals, where he accurately distinguishes the true from the false; describes all the kinds of those that are true, and shows that they were real Hebrew coins struck by the Jews, after the models of the ancient monies, and that they were current before the captivity of Babylon. All those medals however, that we see with the head of Moses and Jesus Christ, are manifestly false. It has been a pious or superstitious fraud, but still more commonly a thirst of gain that has fabricated these. Lastly, it is necessary to observe, that the Jews counted by talents, shekels, bekas, zuzas or dracmons, and by geras. The gera was equal to six sols of France, or three English pence. There were shekels of gold and of silver; the silver shekel is that which is commonly taken for a penny, and of which the Jews gave thirty

to

to Judas as the price of his perfidy in betraying our Saviour. It has on one side the figure of Aaron's rod, with this inscription, Jerouchalaim Hakkedoucha, Jerusalem the holy; and on the other the cup in which the manna was kept, that was preserved in the sanctuary, with these words round it, Chekel Ischrael, or the money of Israel. After the Romans became masters of Palestine, the Jews put the image of the emperors on their coins, as appears by the words of our Saviour himself, in chap. 20. of the gospel of St. Luke.

XI. There are likewise,

2. Egyptian medals, which are very rare.

3. Chinese; but of which there are scarce any that are antique.

4. Syriac.

5. Persian.

6. Arabic.

7. Greek; which are the most beautiful of all: for the Greeks struck coins in all the three metals, with an art so excellent, that the Romans were never able to equal them. The figures on the Greek medals have a design, an attitude, a strength and delicacy, which expresses the muscles and the veins in a manner infinitely superior to those of the Romans. These are very scarce and extremely valuable.

8. The Roman; which are elegant, common, and authentic, and of which a series may be formed

formed almost without any interruption. We shall hereafter speak more fully of these.

9. The Hetruscan; of which it is pretended there are still some to be found, but of this many learned men have a rational doubt.

10. The Punic or Carthaginian: these are not scarce, especially in small brass. They are easily distinguished by their emblem, which is a crocodile resting against a palm tree; and was the arms of the republic of Carthage. There are some of them also that have a human figure on one side holding a spear in one hand, with this inscription Kart-hago; and on the other the head of a horse, in profile, and under, on the exergue is XI$\underline{\text{II}}$.

11. The Parthian.

12. The Gothic: these are ill shaped pieces, and of which neither the characters nor emblems are explicable. The Goths, having made themselves masters of Italy, would imitate the emperors, and caused money to be immediately struck, with a form and character of their own: but they succeeded very badly; and in their gold coins there is not sometimes a fourth part that is pure. There are however some medals of their kings, as Atalaric, Theodal, Witiges, Totilas, Attila, &c. which should be ranged after the last emperors of the West.

13. The Spanish; which were made in imitation of the Punic, because the Carthaginians were then masters of Spain; and they had particular

ticular characters which no body now understands.

14. The modern European.

15. The Miscellaneous: these do not belong to any regular series or system; but have been struck by some particular city. These are met with by chance, and there is frequently much labour required to decypher and explain them.

XII. It would be to enter an immense labyrinth, were we here to attempt to describe all the different coins and medals, and to point out their characteristics. We must content ourselves with explaining their general qualities. Every medal has two sides, which are called its face and reverse. On each side there is the field, which is the middle of the medal; the rim or border; and the exergue, which is that part that is beneath the ground on which the figures stand. On the two sides they distinguish the type, and inscription or legend. The type is the figures that are represented; and the inscription, or legend, is the writing, and principally that which is on the rim. Though frequently in the Greek medals, and sometimes in the Latin, the inscription is on the field. That which is on the exergue is less commonly called inscription, because it frequently consists of initial letters only, the meaning of which is not understood. Those which are counterfeits of the antique are called false medals: those which are in part effaced are named defaced medals: such as

are

are called restored medals have the letters *rest.* on them; which show that they were restored by the emperors, in order to render them perpetual: those that were made of copper, and afterward covered with silver, are called, cased medals: such as have only a very thin coat of silver over the copper, but which are so dextrously done that it cannot be perceived, without cutting them, are said to be plated: cleft medals are those that are cracked on the edge by the force of the stamp: those that are notched on the edge are called indented medals; this is a proof of their value and antiquity: incuse medals are such as have no reverse: counter-marked medals are such as have a stamp either on the face or on the reverse, and which shows that they have changed their value; the curious make much search after these: cast medals are such as were not struck, but cast in a mould.

XIII. To give our readers an idea of the method of examining all sorts of medals; and of making a just and learned decision concerning them, we shall take, as an example, the coins and medals of the Romans, which may serve as models in every respect; and of which we have remaining the most complete series. They therefore consider,

1. The metal. Gold medals are not liable to be injured by rust; and the gold of which they are made is very pure, even finer than that of the Holland ducats. There is no
great

great number of their silver medals; and they are commonly very small: of these there can be no remarkable class or division formed: silver is likewise not subject to rust. The medals of brass, and the coins of copper, are, on the contrary, so numerous, that a regular and complete system may be formed of them. They are divided, as we have said, into large, middle and small. The connoisseurs pretend also that there are likewise some of Corinthian metal. There is found on the copper an antique rust, that resembles a varnish, and is called patima: it is of a variable colour between green and black, and prevents the rust from eating any further. This rust the moderns have not hitherto been able to imitate. There are also medallions that are called *Æris moduli maximi*, and *Æris maximi*; and which are known by not having the usual mark of the letters S. O. There are also medals or coins of iron, tin, and even lead *(plumbei nummi.)*

XIV. 2. The time when they were struck. In the Roman medals they distinguish two periods. The first is of those that were made in the time of the republic, and are named *Nummi consulares*: and the second is of those that were struck under the emperors, and are named *Nummi imperatorum*, and *Imperiales*.

3. The representation of a medal: in which they examine, 1. on one side the face, image or bust: 2. the reverse, or iconologic representation:

tion: and here we must observe, that iconology is the art of representing to the sight all sorts of memorable events by images or symbols, in which a corporeal figure represents a moral or ideal object. The Greeks and Romans made frequent use of these in their medals. And 3. the inscription; in which the ancients employed particular abbreviations, that are necessary to be known: thus S. F. signified *saeculi felicitas*: T. F. *temporum felicitas*: C. R. *claritas reip*: S. A. *spes Augusta*, &c. Sometimes also the name of the city is seen, where the piece was made; or monograms, with the name of the master of the mint, and other like matters.

XV. 4. The value of the coin, as it passed in the community where it was made; such as the *denarii, asses, quinarii, sestertii, sesquitertii, libella, simbella*, &c. These values are commonly marked on the coins by signs.

5. The singularity or scarcity of a medal, which forms its hypothetic value. Thus, in the Roman medals, those of Otho, Pertinax, Gordianus, Africanus, are of inestimable worth, because they are, so to say, singular. In like manner when there are two busts or heads together *(capita jugata)* &c. with other important or remarkable singularities.

6. The preservation of a medal; that is, whether it has been well preserved, or effaced, diminished, or injured by rust, or otherwise damaged; which diminishes its value. There are

are sometimes medals found so fair and fresh, that they appear as if they were just come out of the mint. The French name these a *Fleur de coin*, and the Italians, *Di tutta offervazione*.

7. The beauty of the design, and the perfection of the engraving, as well as the *relief*, in which the ancients, and especially the Greeks, excelled. By this is meant the whole composition of a medal. To judge properly of these matters, it is necessary to understand drawing, and engraving; to be a connoisseur in the polite arts; and, by seeing a great number of such as are excellent, to form a refined taste.

8. In the consular coins or medals, the Roman family to which they belong is also to be considered. There are medals of 178 illustrious families of Rome.

9. Lastly, in order to understand well, and properly judge of antique coins, we should be versed in history and antiquities, and know the customs, ceremonies and manners of ancient nations.

XVI. But as the medals of the ancients have been frequently counterfeited, and as it is of great consequence not to be deceived in this matter, numismatography points out to us the principal characteristics of these counterfeits, and the marks by which we may distinguish the true from the false. These fictitious medals are therefore divided into five classes. 1. Such as have been designed and made, in modern times, in imitation of thofe of the antique. 2. Such as have been accurately copied after fome antique medal that really exifts. 3. Thofe that have been formed or caft in the mold of an ancient medal. 4. Such as are compofed of two antique medals, by cementing or joining them together. 5. Thofe that are really antique, but that have been altered and fophifticated. But notwithftanding all the precautions that numifmatography gives in full detail, it is ftill very difficult for the connoiffeur to avoid thofe fnares that are continually laid for him; and even not to be frequently deceived.

XVII. With regard to the methods of which the ancients made ufe in forming or coining their moneys and medals, we know but little of the matter. The opinions of the learned concerning it differ widely. *Ottavio Ligorio*, an *Italian* antiquary, imagines that they drew the defign on the medal itfelf, and afterward graved it in relief. To conclude; the moft celebrated writers on medals are, *Antonius Auguftinus*, bifhop of *Tarracon*; *Wolff*, *Lazius*, *Fulvius Urfinus*, *Hubertus Goltzius*, *Andrew Schot*, a jefuit, *Lewis Nonnius*, a phyfician, *Æneas Vicus*, *Oifelius*, *Seguin*, *Occo*, *Triftan*, *Sirmond*, *Vaillant*, *Charles Patin*, *Noris*, *Spanheim*, *Hardoin*, *Morel*, *Joubert*, count *Mezzabarba*, *M. Begber*, &c. Father *Bandouri* has placed, at the head of his collection of medals, *Bibliotheca nummaria, five auctorum qui de numifmatibus fcripferunt*.

CHAP.

CHAP. XII.

DIPLOMATICS.

THIS science does not, nor can it, extend its researches to antiquity; but is confined to the middle age, and the first centuries of modern times. For though the ancients were accustomed to reduce their contracts and treaties into writing, yet they graved them on tables, or covered them over with wax, or brass, copper, stone or wood, &c. And all that in the first ages were not traced on brass or marble has perished by the length of time, and the number of destructive events. Notwithstanding which, diplomatics must not be regarded as a trifling science, or as of mere curiosity: on the contrary, it is useful, indispensable, and of the greatest importance to erudition in general, and to literature in particular.

II. As the objects which enter into diplomatics, and on which it is exercised, make it a distinct science, it is therefore only necessary to know those objects and their denominations, as they have been described by the learned of different ages. We shall begin by explaining the peculiar terms of the art; and we imagine that

it

it will be afterward eafy to explain the fyftem of the fcience itfelf.

III. The word diploma fignifies, properly, a letter or epiftle, that is folded in the middle, and that is not open. But, in more modern times, the title has been given to all ancient epiftles, letters, literary monuments, and public documents, and to all thofe pieces of writing which the ancients called Syngrapha, Chirographa, Codicilli, &c. In the middle age, and in the diplomas themfelves, thefe writings are called Litteræ, Præcepta, Placita, Chartæ indicula, Sigilla, and Bullæ; as alfo Panchartæ, Pantochartæ, Traƈtoriæ, Defcriptiones, &c. The originals of thefe pieces are named Examplaria, or Autographa, Chartæ authenticæ, Originalia, &c. and the copies, Apographa, Copiæ, Particulæ, and fo forth. The colleƈtions, that have been made of them, are called Chartaria and Chartulia. The place where thefe papers and documents were kept, the ancients named Scrinia, Tabularium, or Ærarium, words that were derived from the tables of brafs, and according to the Greek idiom, Archeium or Archivum.

IV. In order to underftand the nature of thefe ancient papers, diplomas and manufcripts, and to diftinguifh the authentic from the counterfeit, it is neceffary to know that the paper of the ancients came from Egypt, and was formed ed of thin leaves or membranes, taken from the branches of a tree, named Papyrus, or Biblum Ægypthiacum, and which were pasted one over the other with the slime of the Nile, and were pressed and polished with a pumice stone. This paper was very scarce, and it was of various qualities, forms and prices, which they distinguished by the names of charta hieratica, luria, augusta, amphitheatrica, saitica, tanirica, emporetica, &c. They cut this paper into square leaves, which they pasted one to the other, in order to make rolls of them; from whence an intire book was called volumen, from volvendo; and the leaves, of which it consisted, paginæ. Sometimes, also, they pasted the leaves altogether, by one of their extremities, as is now practised in binding; by this method they formed the back of a book, and these the learned call codices. They rolled the volume round a stick, which they named umbilicus, and the two ends, that came out beyond the paper, cornua. The title, wrote on parchment, in purple characters, was joined to the last sheet, and served it as a cover. They made use of all sorts of strings or ribbands, and even sometimes of locks, to close the book, and sometimes also it was put into a case. But there is not now to be found, in any library or cabinet whatever, any one of these volumes. We have been assured, however, by a traveller, that he had seen several of them in the ruins of Herculaneum, but so damaged, the paper so stiff and brittle,

tle, by the length of time, that it was impossible to unrol them, and consequently to make any use of them, for on the first touch they fell into shatters. We shall speak hereafter of those books they call codices.

V. We are ignorant of the precise time when our modern paper was invented, and when they began to make use of pens in writing, instead of the stalks of reeds. The ink, that the ancients used, was not made of vitriol and galls, like the modern, but of soot. Sometimes also, they wrote with red ink, made of vermilion, or in letters of gold, on purple or violet parchment. It is not difficult for those, who apply themselves to this study to distinguish the parchment of the ancients from that of the moderns, as well as their ink and various exterior characters: but that, which best distinguishes the original from the counterfeit, is the writing or character itself, which is so distinctly different from one century to another, that we may tell with certainty, within about 40 or 50 years, when any diploma was written. There are two works which furnish the clearest lights on this matter, and which may serve as sure guides in the judgments we may have occasion to make on what are called ancient diplomas. The one is the celebrated treatise on the Diplomatic, by F. Mabillon; and the other, the first volume of the Chronicon Gotvicense. We there find specimens of all the characters, the flourishes,

ishes, and different methods of writing of every age. For these matters, therefore, we must refer our readers to those authors; and shall here only add, that,

VI. All the diplomas are wrote in Latin, and consequently the letters and characters have a resemblance to each other; but there are certain strokes of the pen, which distinguish not only the ages, but also the different nations: as the writings of the Lombards, French, Saxon, &c. The letters in the diplomas are also usually longer, and not so strong as those of manuscripts. There has been also introduced a kind of court hand, of a very disproportionate length, and the letters of which are called *Exiles litteræ, crispæ ac protractiores*. The first line of the diploma, the signature of the sovereign, that of the chancellor, notary, &c. are usually wrote in this character.

VII. The signature of a diploma consists either of the sign of the cross, or of a monogram or cypher, composed of the letters of the names of those who subscribed it. The initial letters of the name, and sometimes also the titles, were placed about this cross. By degrees, the custom changed, and they invented other marks; as for example, the sign of Charlemagne was thus:

$$\mathrm{x} \xrightarrow{\overset{R}{\wedge}} \mathrm{s}$$
$$\underset{L}{\vee}$$

They sometimes added also the dates and epoch of the signature, the feasts of the church, the days of the calendar, and other like matters. The successive corruption of the Latin language, the style and orthography of each age, as well as their different titles and forms; the abbreviations, accentuation, and punctuation, and the various methods of writing the dipthongs, all these matters united, form so many characters and marks, by which the authenticity of a diploma is to be known.

VIII. The seal, annexed to a diploma, was anciently of white wax, and artfully imprinted on the parchment itself. It was afterward pendent from the paper, and inclosed in a box or case, which they called Bulla. There are some also that are stamped on metal, and even on pure gold. When a diploma bears all the characters that are requisite to the time and place where it is supposed to be written, its authenticity is not to be doubted; but, at the same time, we cannot examine them too scrupulously, seeing that the monks and priests, of former ages, have been very adroit in making of counterfeits; and the more, as they enjoyed the confidence of princes and statesmen, and were even sometimes in possession of their rings or seals.

IX. With

IX. With regard to manuscripts that were wrote before the invention of printing, it is necessary, 1. to know their nature, their essential qualities and matter; 2. to be able to read them freely, and without error; 3. to judge of their antiquity by those characters which we have just mentioned with regard to the diplomas; and 4. to render them of use in the sciences. As there are scarce any of the ancient codes now remaining, (see sect. IV.) wrote on the Egyptian paper, or on wood, ivory, &c. we have only to consider those that are written on parchment or vellum (membraneos) and such as are wrote on our paper (chartaceos). The former of these are in most esteem. With regard to the character, these codes are written either in square and capital letters, or in half square, or round and small letters. Those of the first kind are the most ancient. There are no intervals between the words, no letters different from the others at the beginning of any word, no points, nor any other distinction. The codes, which are wrote in letters that are half square, resemble those we have in Gothic characters, as well for the age, as the form of the letters. Such as are wrote in round letters are not so ancient as the former, and do not go higher than the ninth or tenth century. These have spaces between the words, and some punctuation. They are likewise not so well wrote as the preceding, and are frequently disfigured with comments. The codes are divided, according to the country, into Lombard,

bard, Italian, Gaulic, Franco-Gaulic, Saxon, Anglo-Saxon, &c.

K. In the ancient Greek books, they frequently terminated the periods of a difcourfe, inftead of all other divifion, by lines; and thefe divifions were called, in Latin, *verfus*, from *vertendo*: for which reafon thefe lines are ftill more properly named *verfus* than *lineæ*. At the end of a work, they put down the number of verfes of which it confifted, that the copies might be more eafily collated: and it is in this fenfe we are to underftand Trebonius, when he fays, that the pandects contain 150000 *pæne verfuum*. Thefe codes were likewife *vel probæ vel deterioris notæ*, more or lefs perfect, not only with regard to the calligraphy or beauty of the character, but to the correction of the text alfo.

XI. It is likewife neceffary to obferve, in ancient codes, the abreviations, as they have been ufed in different centuries. Thus for example, A. C. D. fignifies, Aulus Caius Decimus; Ap. Cn. Appius Cnaius. Aug. 'mp. Auguftus Imperator. The characters, that are called *notæ*, are fuch as are not to be found in the alphabet, but which, notwithftanding, fignify certain words. All thefe matters are explained in a copious manner by Voffius, and in the Chronicon Gotvicenfe. Laftly, the learned divide all the ancient codes into *codices minus raros, rariores, editos*

editos & anecdotos. The critical art is here indispensably necessary; its researches, moreover, have no bounds, and the more, as the use of it augments every day, by the discoveries that are made in languages, and by the increase of erudition.

XII. We might here speak of the invention of printing, and of the different characters of books that have appeared since that epoch: but all that concerns printed books, seems to appertain less to the diplomatic, which relates to manuscripts, than to the knowledge of authors; we shall therefore take due care, when we treat on that part of literature, to mention every thing material that relates to the art of printing.

CHAP. XIII.

STATISTICS.

I. AFTER having learned the ancient state of the world by history, by antiquities, medals, and the diplomatic art, it is both natural

tural and juſt, to deſire to have a knowledge of the ſtate of the preſent world, and of the moſt important occurrences of our own days; and this we learn by Statiſtics, by the relations of travellers, and by geography. The ſcience, that is called *Statiſtics, teaches us what is the political arrangement of all the modern ſtates of the known world.* This arrangement, comprehended formerly under the title of the political ſyſtem, has been known and explained very imperfectly, not only with regard to diſtant and ſmall ſtates, but even large kingdoms, ſituate in the center of Europe. In geographical treatiſes, they placed, before the local deſcription of each country, a ſort of account of the principal objects that compoſed its ſyſtem. But theſe introductions were always imperfect, naturally very contracted, frequently dubious, and ſometimes abſolutely falſe, or ill grounded. We muſt except ſome of them however, eſpecially thoſe which are to be found in the excellent geography of M. Buſching, an author, whoſe aſſiduity, preciſion, and diſcernment, can never be ſufficiently commended. But this book has, as we may ſay, but juſt appeared in its full perfection.

II. The hiſtorians have not been leſs ſenſible of the neceſſity of making their readers acquainted with the political ſyſtem of the principal modern ſtates of Europe; and the celebrated Baron Puffendorff, in his univerſal hiſtory, has annexed

annexed, to that of each country, an abridged relation, which contains some instructions relative to this matter. But 1. these sort of instructions are frequently erroneous, and always imperfect or defective; 2. they are too much dispersed to be used as a systematic abridgement, which might serve as the basis of public or private lectures; 3. the daily occurrences that happen in the world, and especially the treaties of peace, are constantly changing the system of governments, and make the statistic science a kind of moving picture, where the momentary situation of the parts is much better seen in a course made by an able professor, than in a book; which loses its accuracy and use in proportion as it grows old. These considerations, and numberless others, have induced authors of ability to furnish the world with instructive descriptions of this nature.

III. Thus, the Thirty two republics of the Elzeviers, which appeared more than a century since; the work of Frederic Achillis, duke of Wirtemburg, intitled Consultatio de principatu inter provincias Europæ operâ Thomæ Lansii, Tubingæ 1655; Le Monde, by Peter D'Avity; Gothofredi Archontologia cosmica; Lucas de Linda, Descriptio Orbis; Hermanni Conringii, opus posthumum, de notitia Rerumpublicarum hodiernarum; J. C. Beckman, Historia orbis terrarum, geographica & civilis. Many statesmen also have employed themselves
in

in describing some particular states to their cotemporaries; thus toward the end of the sixteenth century there appeared, the relations of some Venetian ambassadors: the embassies of the Earl of Carlisle, an English minister: Molesworth's account of the state of Denmark; and a number of other works of the same kind. M. Everhard Otto, professor at Utrecht, and afterward senator at Bremen, was the first who made a collection of these scattered accounts, and, by adding his own informations, composed a very good work, under the title of Notitia præcipuarum Europæ Rerumpublicarum. We have also La description du monde, de Jean Funck: and a very good work in English, intitled Modern history, or the present state of all nations, by Mr. Salmon, illustrated with cuts, London 1744. This work has been translated into Italian and Dutch, with some advantageous alterations.

IV. It would be far from just, in this place, to pass over in silence the obligations this science has to M. Godfrey Achenwal, professor at Gottingen, who has not only composed an Introduction to the political system of the modern states of Europe; and another work not less interesting, intitled Principles of the history of Europe, leading to the knowledge of the principal states of the present time; but has been also the first to reduce this important

subject

subject into a true system, and has made a separate science of it, under the title of Statistics; and which he professes with great reputation: a science from which history borrows great lights; which furnishes the best materials for the constitution of a state, which enriches politics, and which prepares those of the brightest genius among the studious youth, to become one day able ministers of the state.

V. All that occurs in a state is not worthy of remark, but all that is worthy of remark in a state, enters necessarily into statistics. This science begins therefore by making, 1. An exact division of the four parts of the world, and shows into how many states, nations, monarchies, republics, and lesser governments, each of these parts is divided. It is scarce necessary to observe, that the knowledge of the states which belong to Europe are the most important.

2. It proceeds to the examen of each particular state, and of its revolutions; and here it has an especial regard, 1. to the principal epochs; 2. to the changes that have occurred in the form of government; 3. to the provinces that have been conquered or acquired by a state, or that have been dismembered from it, and 4. to the hereditary governments, and the alterations that have happened in families.

VI. Each

VI. Each state consists of country and inhabitants. Under the title of country, statistics comprehends,

3. The extent of territory in a state, its local situation, the rivers by which it is watered; the sea that washes its coasts, its borders, its mountains, and natural productions. It inquires into the state of its capital; or the seat of government, its exterior possessions, and especially its colonies, in the other parts of the world, &c.

4. With regard to the inhabitants; it inquires into their number and qualities: and for this purpose it makes, by the aid of political arithmetic, of registers of births and burials, &c. the most elaborate and accurate researches possible, into the number of the inhabitants of a state, and into their genius; the prevailing character, the industry, the virtues and vices of a nation.

VII. 5. It next considers the inhabitants under the quality of citizens, united by laws for their common interest; and in this light, the sovereign himself is nothing more than the first citizen. And here it directs its views to two principal objects, which are 1. all that relates to the constitution of a state, and 2. all that enters into the arrangement of its public affairs. It examines, therefore, what are the fundamental laws, the usages and customs received in a country,

try, and which have there the force of laws, &c. From thence it paſſes,

6. To the rights, privileges and prerogatives of kings and other ſovereigns, or of ſenates and magiſtrates: it conſiders the manner of attaining to the throne or government; the limits preſcribed by each country to the authority of its ſovereign, or other governors; and ſo of the reſt.

7. The rights of the ſtates of a nation, of the nobility, clergy, military, citizens, and peaſants; the diets and other public aſſemblies for deliberating on affairs of importance, &c.

VIII. When a ſolid knowledge is acquired of all theſe matters, ſtatiſtics paſſes to the examen of the diſpoſitions eſtabliſhed in each country, for the conducting of public affairs: and it ſhows,

8. The dignity, rank, title, and arms; the court, ceremonial, orders of knighthood, &c. of the ſovereign.

9. The arrangement of the department for foreign affairs, or the cabinet.

10. The diſpoſitions in the direction of interior affairs, for the eccleſiaſtic ſtate, the adminiſtration of juſtice, the finances, commerce, the ſciences, and the military: and here it enters into the following particulars.

IX. 11. It conſiders what is the eſtabliſhed religion of a country, and what other religions
are

are there tolerated; and their several rights, not only as they relate to the state, but with regard to each other. The privileges of each church, the rights of the clergy, the several orders of ecclesiastics, their principal functions, charges, revenues, &c.

12. The laws civil and municipal, the tribunals of justice, the forms of process, and the criminal laws and jurisprudence.

13. The principal regulations with regard to the police.

14. The resources of the state, 1. in its agriculture and all its natural productions; 2. in its manufactures and fabrics; 3. in its commerce interior and exterior, active and passive; and 4. in its mercantile navigation.

15. In the arrangements of its chambers of finances, the domains of princes or states, the royalties, contributions, and all the subsidies that the subjects pay to the sovereign for the support of government: in a word, all the revenues of a state, and the manner of collecting and employing them.

X. Statistics then considers,

16. The state of the arts and sciences, which do so much honour to a nation; what schools, colleges, academies and universities there flourish; what remarkable public libraries they have; what artists there excel; and what encouragement all these receive from the state.

XI. Laftly, as the military ftate is now become a neceffary evil in the political fyftem of modern Europe, this fcience applies itfelf particularly to the defcription of

17. The number of troops that each ftate maintains, the arrangement of the army, what is the difpofition of each people for war, the goodnefs of their troops, their difcipline, their uniform, their arms, the refpective numbers of cavalry and infantry, the ftate of its artillery and arfenals, its fortifications, the facility with which it raifes recruits, its barracks, hofpitals for invalids, its engineers, cadets, and every thing that can have any relation to the military ftate.

18. It confiders, after the fame manner, the marine of a nation, the number of its fhips of the line, frigates, bomb veffels, firefhips, &c. the number and ability of the failors each ftate can furnifh; the arrangement of its docks, yards and arfenals for the marine; the materials for the conftruction, equipment and victualling of fuch fhips as the ftate can furnifh, or as the government is obliged to draw from other parts; the fchools for the marine, and all other objects relative to this article.

XII. The laft inquiry in which ftatiftics is employed, is in explaining what is the true intereft of each nation. Now this intereft is either,

19. Internal;

19. Internal; and relates to the tranquillity, prosperity, and increase of a people, in its industry, its manners and politeness; its riches, refinements and opulence. Or,

20. External; and relates to the maxims of government that are proper for it to observe with regard to its neighbours, its allies, neutral powers, and even with regard to its enemies: maxims which ought to be founded on the local situation of each country; on the rivality either greater or less in commerce; on the apparent views of increase of power that a state may have; on family compacts or consanguinity; on alliances, either perpetual, or limited to a time or an object; on the proportion of power; and on an infinity of similar relations.

XIII. They who teach the statistic science as public professors, or write expressly on this subject, endeavour to explain all these various objects as they regard each nation, country, or particular state. It is true, that they are sometimes mistaken in their conjectures: it is likewise true, that a man of letters is not a minister of state, and frequently a minister of state is not a man of letters: it sometimes happens, however, that, by force of reflection, a man of genius and learning becomes enabled to discover the true interests of a state, especially those that are natural and immutable; while the politician mistakes those transient interests, of which he makes such wonderful mysteries.

XIV. We

XIV. We have remarked in the second section, that the books which treat on statistics, or the descriptions of modern states, which approach nearest the exact truth, are made to recede from it by time, by those vicissitudes to which all human institutions are liable, and which arise as well from the daily occurrences, and from those grand revolutions that are natural to every state. This is an unavoidable inconvenience, and for which there is no remedy but the constant and judicious perusal of the gazettes and political journals, as the Historical Mercury, &c. These daily and periodical publications afford a continual supplement to the best statistic authors, and form a kind of practical statistics. It is for this reason that the German professors make constant use of them in the universities; for in reading the best gazettes that are brought by each post, they explain to their auditors, not only the terms, the facts, and the causes of events, but by applying these facts and events to statistics, they shew the alterations that are thereby caused in the constitution of the country to which they relate. But, to answer this purpose, it is necessary to make use of the best gazettes of the time, that is, such as are esteemed of the greatest veracity, whose authors are not in haste to insert reports which they are afterwards obliged to contradict; and that are not infected with a national partiality, or a predilection for a particular court or party, and that do not load their relations with insipid or malignant reflections,

flections, nor assume the gift of predicting future event; but such as recount, neither too soon, nor too late, the several events as they arise, in a natural style, in a faithful and impartial manner, and without gloss or comment; leaving to their readers the care of making, on each event, their critico-politico-prophetical reflections.

※※※※※※※※※※※※※※※※※

CHAP. XIV.

OF

TRAVELS and TRAVELLERS.

A Great traveller makes a good liar, says the proverb; and Strabo asserts, that every man, who relates his travels, relates falsities: but whatever the proverb or Strabo may say, it is to the relations of travellers that we owe our knowledge of the state of the world, and especially of such countries as are at a great distance from us. The utility of these relations; their great number, which amounts to more than 1300 that are already printed; the satisfaction

faction they afford our curiosity; the assiduity with which men of letters, as well as men of the world, apply to these authors, and many other considerations, have made the study of voyages and travels a considerable branch of Universal Erudition: it appeared therefore necessary to make of it here a distinct chapter.

II. Whenever a man passes from one country or province to another, he is said to travel; but the travels of which we here speak are those that are made into far distant countries, and that are undertaken with various views. We are not here to consider the voyages of merchants or seamen, who traverse the sea from motives of commerce, nor the journeys of such men whose private affairs carry them into distant countries, but we are here to treat of the travels of those whom a desire of knowledge, and of communicating their discoveries to mankind, have induced to undertake long journeys. Thus the indefatigable inquirer, after philosophical knowledge, searches every part of the globe in pursuit of new discoveries in natural history, botany, &c. or descends with his thermometer into the deepest caverns. Thus the sagacious astronomer transports himself, sometimes to the equator, and sometimes to the poles, intent upon making accurate observations on the heavenly bodies; or on measuring the degrees of the earth. Thus the learned antiquary traverses Italy, Greece, Asia Minor, Palestine, Egypt, and

and all the regions of the east, in quest of those precious monuments of antiquity, which may lead to the knowledge of ancient history. Thus the politician visits every civilized nation, in order to learn their manners, their policy, and form of government. And thus the man of curiosity flies to the most distant parts of the earth, in search of unknown nations, and to gratify his desire of making new discoveries. It is, therefore, to these five objects that we may reduce the design of all travellers.

III. It were to be wished, that all, who undertake long journeys from either of these motives, would not only provide themselves with all the necessary preliminary knowledge, which could give them a well grounded hope of success in their attempt, but also, that before they engage in so difficult an enterprise, they would lay down a judicious plan for their journey, and for all the objects that relate to their inquiries. It were also to be wished, that they would communicate their design to the public, at least a year before they set off, by an advertisement in all the literary gazettes, that the learned might be induced to communicate their salutary informations and advice, relative to the undertaking. Whoever has read the instructions that were drawn up by M. Baumgarten, professor at Hall, for the young batchelors of arts, who were sent for to teach philology, in a celebrated Greek convent situate on the promontory of Athos,

Athos, and those which professor Michaëlis of Gottingen gave to the learned men, who were lately sent to the Holy Land, and other parts of Asia, by the king of Denmark, will clearly see the importance, utility, and even indispensable necessity of such informations. He, who does not know what it is he ought to inquire after, can never expect to find, except it be by chance, any thing remarkable that others have not found before him. It were to be wished, in the last place, that no one would undertake such a journey, without the company of some one skillful in drawing, and even in geometry; for there are a thousand occasions where it will be necessary to measure altitudes and distances, and a thousand objects, of which adequate descriptions cannot be given, of which we cannot form a true idea, without the help of figures.

IV. During the course of his journey, the traveller cannot be too much on his guard, as well against his own credulity, as the snares that will be laid for him by the inhabitants of the countries through which he shall travel. All nations of the earth, and especially those of the warm climates, are full of ancient traditions and fables; which, if he should believe, would carry him far distant from the truth. Herodotus, Diodorus Siculus, and almost all the ancient historians, geographers, and travellers, have been the dupes of these relations. We cannot read, without disgust, the idle tales they recount,

and

and by which their wretched credulity is inceſſantly ſhown. We are inclined to ſay to the travellers who relate ſuch tales: *Sir, if I had ſeen theſe things myſelf, I ſhould not have believed them, but I believe them becauſe you have ſeen them.* A traveller ſhould examine all things with his own eyes, and ſhould write down all he ſees on the ſpot, in his protocol, or itinerary. Idleneſs is incompatible with accuracy, and whoever is fearful of inconvenience, fatigue and expence, muſt never hope to produce a relation of his travels worthy of regard.

V. A deſire of recounting marvellous relations is natural to all travellers; but they ſhould remember, that all that is marvellous appears conſtantly ſuſpicious to a rational mind; and that it is even more prudent to ſuppreſs facts, which, though true, are incredible, than to render their veracity doubtful by aſtoniſhing reports. Candor, ſincerity, accuracy, and a judicious diſcernment, ſhould be conſtantly conſpicuous in every relation. The ground work ſhould be laid in truth, and the ornaments ſhould be pleaſing and judicious: for it is by juſt and pertinent reflections, that relations of this kind are prevented from becoming dry and diſagreeable.

VI. There are but few good relations of travels made in Europe; becauſe it has been very difficult, and even dangerous to ſpeak the truth.

It

It should seem as if the people were ashamed of their countries, and the princes of their conduct. Such as have given true accounts have been persecuted for their veracity. The travels of Keisler, in Europe, are the most esteemed, and the most worthy of estimation. There are some made in other parts of the world, that are very valuable. The travels of Tavernier in Turkey, Persia, Mogustan, &c. are much celebrated, but the strict truth does not always appear manifest in them: the method of valuing diamonds according to their size and weight, and the perfection of the water, is the most interesting article they contain. The travels of Chardin in Persia, of DuHalde in China, of Kæmpfer in Japan, of Shaw in Egypt, of Kolbu to the Cape of Good Hope, the relation which M. de la Condamine made to the academy of sciences on his return from America, the celebrated voyage of Lord Anson round the world, &c. are master-pieces of this kind, and may serve as models to all who shall hereafter undertake similar enterprises.

VII. We owe to England the first idea of an admirable work, consisting of a vast collection of the best relations of travels and voyages, and reduced into a regular system. This work first appeared at London under the title of a Collection of voyages and travels, in folio; the first four volumes in 1704, and the fifth and sixth in 1732, and the seventh and eighth in 1747.

1747. This grand work has been tranflated into almoft all languages, but particularly into German, and French by abbé Prevot, under the title of A General Hiftory of Voyages and travels, in thirteen quarto volumes, Paris 1744, and at the Hague 1746. The French tranflation, but efpecially the German, is enriched with many notes that are inftructive, and that rectify confiderable errors in the original. Whoever has courage enough to attempt, and perfeverance enough to labour through thirteen quarto volumes, may acquire a complete knowledge of all travels that have been undertaken, and of all the known countries in the four parts of the globe, without having fcarce any occafion to apply to other books of the kind. This work may however, at all times, be of ufe as a dictionary, to be confulted occafionally concerning any particular country of which we may want information.

VIII. In a kingdom that is furrounded by the the fea, and whofe power arifes from navigation, it appeared neceffary to render thefe accounts interefting to mariners. For which reafon there are many matters inferted which appear to be calculated merely for them, as accounts of foundings, of rocks, coafts, of the entrance of harbours, of trade and variable winds, &c. But every reader who is not interefted in navigation, may eafily pafs over thefe tedious articles.

IX. We

IX. We should be liable to be equally tedious, were we to attempt to inform such as may undertake what is called a literary journey, of all the objects that ought to attract their curiosity: of the most celebrated among the literati, whose acquaintance they should endeavour to acquire: of the public and private libraries they should visit, as well as the cabinets of natural history, antiquities, medals, coins, paintings and other curiosities: the monuments of every kind they should examine: the observations they should make relative to the character, the genius, humanity, and politness of each nation: on the different forms of government: on the state of letters in each country, its universities, colleges, academies, and an infinity of like matters; as rare manuscripts, remarkable inscriptions, &c. Some learned men have given instructions in form relative to these matters, and among others M. Kohler, a celebrated professor at Gottingen, to whom the world is indebted for many other valuable works.

X. We shall finish this chapter with one remark. Credulity is the source of most errors, as doubt is the beginning of wisdom. It is therefore allowable to entertain a rational pyrrhonism concerning the relations of most travellers, and it is of the last importance to make a judicious choice of such as we propose to read: for the first accounts of any country, or people,
make

make the strongest impressions on our minds, and if they should be false or erroneous, it is almost impossible for us totally to eradicate such impressions, but we shall continue to entertain these false ideas during the remainder of our lives. It is highly necessary, therefore, to be previously acquainted with the degree of reputation each writer of travels bears, for veracity, and for a judicious relation of facts.

CHAP. XV.

GEOGRAPHY.

THE world swarms with descriptions of the world: they appear as if they sprung from the earth like vegetables. There are to be found, in all languages, copious, complete, abridged, systematic and universal geographies; elements, introductions, essays and dictionaries of geography; with numberless other like works. This science is taught in schools, academies, universities, &c. Professors of geography travel

the

the countries, and teach it to the youth of each town or village through which they pafs. The printfellers fhops are loaded with maps, and the walls of each houfe are covered with them. No branch of learning feems to be fo familiar to mankind as geography: and we fhould therefore be inclined to fupprefs the analyfis of this fcience, if it did not form an effential article in the fyftem of univerfal erudition, and if we did not hope to mention fome matters relative to it that are not very commonly known.

II. Geography is a fcience that teaches the knowledge of the terreftrial globe, or of the furface of the earth; of the fituation of countries, cities, rivers, feas, &c. with the defcription of each of them. There are here fome preliminary and effential diftinctions to be made.

(1.) As our globe forms only a part of the univerfe, geography in like manner makes only a part of cofmography.

(2.) It is the bufinefs of geography to inform us of the fituation and natural productions of the earth in each country or climate, which is alfo called phyfical geography. The civil and political arrangement of ftates or governments does not properly belong to it, that rather appertains to ftatiftics; though many of the beft modern geographers have happily united thefe two branches, by calling the latter political geography.

(3.) Geography is either mathematic or natural. The former confiders the earth in the
same

same manner it does the other celestial bodies; examines its dimensions, its figure and situation in the universe; and, in a word, all that has any relation to the mathematics. As we have sufficiently explained this part of geography in the forty-ninth chapter of the first book, from section seventy-nine to eighty-six, we shall confine ourselves here to natural and physical geography; having also explained what relates to the political part in the chapter on statistics in this volume.

(4.) The knowledge of maps and charts, and the manner of using them, makes also a part of geography.

III. (5.) Geography is likewise either sacred or profane. The former furnishes instructions relative to the peregrinations of the patriarchs, and the travels of the Israelites. It elucidates the predictions of the prophets against certain kingdoms and nations; the wars of the Jews; the travels of St. Paul and the other apostles; the establishment of the church in all parts of the known world, &c. Profane geography is divided into,

(6.) The geography of the ancient and middle ages, and of modern times. Each of these parts comprehends a description of the earth and its various inhabitants, in their proper periods. By the labours of ancient geographers, and the modern authors of maps, we have now a complete atlas of the state of the ancient world.

Vol. III.   T   (7.) The

(7.) The description of any distinct country or region is called chorography.

(8.) Topography is a particular description of any place, in so exact and minute a manner, that no one circumstance relative to it is omitted.

(9.) Hydrography is, lastly, the description of waters; for there are charts that contain merely the plans of those seas, rivers, streams or lakes, by which a country is watered.

IV. As the surface of our globe is divided into land and water, geography makes use of certain terms in describing each of these, of which it is necessary to give here a brief explanation, in order to facilitate the understanding of what we have further to say on this subject.

A continent is a large portion of the earth, that contains several contiguous countries, and that is not surrounded by the sea.

An island is a part of the earth that is surrounded by water.

A peninsula, called in Greek chersonesos, is a piece of land that is almost surrounded by the sea.

An isthmus is a narrow neck of land that joins a peninsula to the continent, as the isthmus of Corinth, Panama, &c.

A defile is a narrow and difficult passage between mountains.

A strand is a flat and sandy shore, which the flux and reflux of the sea covers and leaves uncovered each tide.

A

A promontory is a high land that juts into the sea.

A cape is a mountain that in like manner runs into the sea.

A point, on the contrary, is a flat land whose extremity is in the sea.

Downs are small sand hills near the shore.

A beach is a high and steep hill on the shore.

V. With regard to the terms that relate to the water:

An archipelago is a portion of the sea in which there are many islands.

A gulf, or bay, is a part of the sea that runs in between lands.

A strait is a neck of the sea inclosed by two lands, and by which we may pass from one sea to another; it is likewise called a bosphorus, channel, or arm of the sea.

A road is a place proper for casting the anchor, and where ships can ride secure from the wind.

A conflux is that place where two or more rivers join each other.

The mouth of a river is that part where it leaves its bed and runs into the sea, or a lake.

A canal is an artificial river, like those of Ladoga, Languedoc, &c.

A parage is a part of the sea under any given latitude.

VI. In continuing to treat of geography, therefore, we are to be understood to speak of the natural and not the mathematical part, and we mention

mention this once for all. This science begins by examining the properties of the atmosphere that surrounds this globe, the air that we breathe, and the clouds that pass over our heads; the causes of rain, snow, dews, tempests, but especially of the winds, as well those called variable, as the trade winds; as also of whirlwinds and other meteors. It shows that an air charged with vapours is heavier than a clear air*, and consequently more elastic; that it presses more, and that from thence arises that agitation, that motion in the air which is called wind: and that the swiftest wind does not pass over more than fifty feet in a second. It inquires likewise into the causes of the variation of the weather, and the different temperature of each climate.

VII. Geography then proceeds to the contemplation of the earth itself. It examines its mountains and vallies: it considers that chain of mountains of 188 geographic leagues in length, which the Greek and Latin authors call the Alps, and which separate Italy from Germany, and Switzerland from France; those celebrated mountains in South America called the Cordeliers, the highest in the known world, and of which the greatest, named Chimborasso, is 19320 feet above the surface of the sea. It describes the vlocanos

---

* The more general opinion is, that the driest air is the heaviest; and the observations on the barometer seem to confirm this opinion.

of Vesuvius, Ætna, Stromboli, of Hockla and Krabla in Iceland, of their eruptions, their lava, and their effects. It treats likewise of the desarts of those uninhabited countries in northern Asia, which are called Steps; of the natural productions of each climate and country, and of all that relates to the philosophical state of our globe. It then extends its inquiries to the inhabitants of the earth, and endeavours to determine their number, and the principal alterations that attend it, by the aid of political arithmetic: and from thence it concludes, that this earth is capable of maintaining 3000 millions of inhabitants, but that there are not in fact more than 1000 millions existing. It generally allows thirty-three years to each generation: and on that supposition there are 1000 millions of mankind that are born and die within the space of thirty-three years; more than thirty millions each year, 82000 each day, 3400 each hour, 60 each minute, and one each moment. The number of the two sexes is nearly equal, which proves that polygamy cannot contribute to the increase of the human race, and that the celibacy of the clergy, the monks and nuns, is an unnatural and horrid practice. Mankind are distinguished into white, black and mulattoes.

VIII. That part of geography which is called *Hydrography*, or more properly *Hydrology*, examines, in an historical manner, the nature and properties of the water; the sources from whence proceed

proceed those streams that uniting form rivers, which, flowing with different rapidities, sometimes form cataracts, and at last pour their waters into the sea: and it shows that the sea covers near two-thirds of the globe, and bears different names in different regions: the bed of the sea is only a continuation of the surface of the earth, and has like it various inequalities, heights and depths, mountains and vallies, rocks, &c. Hydrology considers also the nature of the waters of the sea, which is more or less salt or bitter in different parts; the motions of its waves, its continual course from East to West, its currents and tides, its gulfs, whirlpools, and fathomless depths.

IX. After these general considerations, geography passes to the examen of the four parts of the world. The earth is divided, 1. Into the old world, which comprehends the three parts that were known to the ancients, Asia, Africa and Europe; 2. The new world, that is America, and 3. The unknown world, as the Terra Australis, and other countries that have not hitherto been penetrated by travellers. The earth has been also divided according to the different shadows: thus the inhabitants of the frigid zones are called Periscii; those of the temperate zones, Heteroscii; of the torrid zone, Amphiscii; and they who have no shadow at noon-day, the sun being directly in their zenith, Ascii. We must here observe by the way, that geographers regard in their operations the north,

north, and that pole, as by that they determine the latitude of places: and the aftronomers obferve the fouth, becaufe from thence they determine the meridian height of the fun and ftars; and it is in that part they obferve the courfe of the zodiac. Another divifion of the earth is that by climates: thus they make twenty-four climates of hours, begining at the equator, proceeding by the degrees of latitude, and ending at fifty-fix degrees thirty-one minutes. They likewife diftinguifh fix climates of days, towards the north, the firft of which begins at the fame degree of fixty-fix, and ends at the pole, where the day is of fix months continuance: thefe latter climates include countries inhabited and uninhabited.

X. But the moft natural divifion, and that which is the moft eafy to be conceived and retained in the memory, is that by which the earth is divided into four parts. Each of thefe four parts is fubdivided into continent and iflands, and geography, by ftill further extending thefe divifions, confiders the ftates or nations that inhabit the feveral parts of the continent and ifles. Thus,

(1.) Europe comprehends 1. toward the north, Denmark, Sweden, Norway, Lapland, Ruffia, including Livonia, Eftonia and Finland, Courland, Pruffia, and Poland with Lithuania: 2. toward the center, that is on the eaft and weft, France, Savoy, Switzerland, Flanders, Holland, Germany,

Germany, including Bohemia, Hungary, Transylvania, Walachia, Moldavia, and part of Tartary: 3. towards the south, Spain, Portugal, Italy, Ragusa, Morea, and Turkey in Europe. The islands that make part of Europe are, 1. in the ocean, Great Britain, including England and Scotland, with the Orcades, &c. Ireland, Iceland, and the isles of the Baltic Sea: 2. in the Mediterranean, Sicily, Sardinia, Corfu, Zante, Cefalonia, Candia, Corsica, Malta, Cerigo, and the islands in the Archipelago, Majorca and Minorca.

XI. (2.) Asia contains Turkey in Asia, Tartary, Siberia, the provinces of the Russian empire in Asia, China, India, Persia, Arabia, and all the provinces and kingdoms that are comprised under those general denominations. The islands that appertain to Asia are 1. in the ocean, the Maldives, Ceylon, Sumatra, Java, Borneo, Ormus, Celebes, the Molucca isles, the Philippines, the Latrones, and the islands of Japan: 2. in the Mediterranean, Cyprus, Rhodes, and some isles of the Archipelago on the coast of Natolia.

XII. (3.) Africa comprehends 1. on this side the equator, Egypt, Barbary, Biledulgerig, the deserts of Zaara, Nigritia, Guinea and Nubia; 2. under the equator, Congo, Ethiopia, in which is Abyssinia; and 3. beyond the equator, the kingdom of Angola, Momemugi, Monomotapa,

nomotapa, Cafraria, Mozambique, Zanguebar, the kingdom of Melinda, the country of the Hottentots, and the cape of Good Hope. The iflands that belong to Africa, and fituate in the ocean are, the Canaries, the ifles of Cape Verd, St. Thomas, the Afcenfion, St. Helena, and Madagafcar.

XIII. (4.) America contains in its continent, which is divided into north and fouth; 1. in the northern part, Nova Scotia, New England, New York, Pennfilvania, Maryland, Virginia, Carolina, Georgia, Florida, Canada, and Mexico or New Spain; 2. in South America are, Terra Firma, Peru, Chili, Patagonia and Paragua, and Brazil. The Dutch have eftablifhments in America at Curacoa, Surinam, and St. Euftatia. The iflands of America are 1. in the eaft, the Azores or the Flemifh iflands, the Antilles, Lucayas, and the Bermudas : 2. in the weft, California.

XIV. (5.) The unknown world confifts 1. under the Artic pole, of the country of Jerfo, New Denmark, New Wales, Labrador, Cumberland's Bay, Greenland, Spitzberg and Zembra or Zembla; 2. under the Antartic pole, Terra del Fuego, New Holland, Los Capous, New Guinea, the iflands of Solomon, la terre de Guis, New Zealand, and all that is included in the Terra Auftralis.

XV. Geography then describes the ocean, and assigns the proper names to the several seas that wash the coasts of the four quarters of the known and unknown parts of the earth; as also the rivers that water each country, and the lakes that they contain. It recounts all the observations that have been made on these seas by mariners, and by naturalists; the productions of each sea; and every other particular relative to them.

XVI. After these general matters, geography passes to the analysis of the several parts of the world: and here it examines their situation and extent; their apparent number of inhabitants, with their figures, qualities, customs and manners; the principal productions of each part of the world; and lastly, the countries and provinces of which it is composed. Each country also undergoes a particular and minute examination, with regard to its natural and political situation, its frontiers and limits, its neighbours and form of government; its capital and other cities, which are divided into great, middling, and small; its sea-ports, fortified places, remarkable towns, palaces, castles, seats, and houses of pleasure; its parks, forests, mines, salt-pits; and, in a word, every object by which it is distinguished or rendered remarkable. From all this it appears, that, to understand geography, it is only necessary to have a good sight, a clear discernment, and a strong memory.

XVII. As

XVII. As geography cannot be well understood without having the objects before our eyes, it is apparent that good maps and charts are indispensably necessary to this science; and as it is necessary to comprehend and remember what we see, it is therefore equally necessary to have complete treatises, as well as abridgments, on this subject. It is not known who was the original inventor of the globe or sphere. John Albert Fabricius has collected, in his Greek Bibliotheque, l. iv. c. 14. the names of those authors who have treated on the globes; and D. Hauber, a German, has given the history of maps. If it be true, that the two globes or balls, in Solomon's temple, were astronomic or geographic globes, they are doubtless the most ancient of which we have any account. According to Diodorus Siculus, Atlas, king of Mauritania, was the first who invented a sphere; which gave rise to the story that Atlas supported the heavens on his shoulders, and was transformed into a mountain. Among the moderns we know of none before those made by Martin Behaim of Nuremberg and Jerome Fracastor. Since their time they have been made by de Hond, Bleau, Coronelli, Gerard Valck, de L'Isle, Moll, Weigel, Beyer, Andreæ, Doppelmayer, Puschner, Lowits, and many other celebrated geographers. There have been some globes constructed of full twelve feet in diameter.

XVIII. With

XVIII. With regard to maps, which form what may be called plans of the earth's surface, they represent 1. either the two hemispheres of the globe; or 2. the four parts of the world; or 3. particular districts; or 4. entire countries; or 5. provinces; or 6. cities and their environs. Charts, on the other hand, represent the different seas, coasts, sand-banks, rocks, &c. They also mark the different depths of the several soundings, the currents, whirlpools, trade and variable winds in each region; the degrees of latitude and longitude, &c. A complete collection of these land and sea charts or maps is called an Atlas. The inventor of maps is no better known than that of globes. Eustathius relates, that Sesostris caused a map to be made of all the countries he traversed; which must certainly be the most ancient. They were also in use among the Greeks and Romans, and other ancient nations. Agathodæmon drew the maps for the geography of Ptolemy, which have come down to us; as well as the famous table of Peutinger that was discovered by Conrad Celtis, purchased by Conrad Peutinger, a nobleman of Augsburg, explained by Beatus Rhenanus, and published by Mark Velserus. After the re-establishment of letters in the sixteenth century, they began again to make maps. Those, which were found in the manuscript of Ptolomy's geography, were the originals of all that have appeared since. Sebastian Munster made them the models of those he designed:

others

others imitated him, and drew maps of particular countries. Abraham Ortelius and Daniel Cellarius collected them, and Gerard Mercator reduced them into a regular system. William and John Bleau, and John Janſſon or Janſenius followed this system. Sometime after, Sanſon deſigned new maps: Francis de Witt and the younger Viſcher improved them, and the Germans copied them; but at length H. Moll, an Engliſhman, and William de L'Iſle, a Frenchman, deſigned and executed maps that were ſo correct and beautiful as to efface the merit of all that had been done before. There is a collection of forty two maps of M. de L'Iſle, that is beheld with admiration by all connoiſſeurs. But as the arts are to be brought to perfection by degrees only, Meſſ. Thomas Kitchin, and J. M. Haſe, have ſtill corrected ſome little inaccuracies in the maps of Meſſ. Moll and de L'Iſle. The coſmographical ſociety of Nuremberg, the academy of ſciences of Berlin, the ſucceſſors of Homarin. Meſſ. Zurner, Scutter, &c. in Germany, Meſſ. Anville, Buache and Bellin in France, and many other able geographers, labour inceſſantly, in giving to maps and charts the greateſt degree of perfection poſſible.

XIX. The beſt maps and charts (and perhaps ſuch only as deſerve to be called good) are thoſe where the ſituation of places and the limits of countries are determined by accurate aſtronomical

astronomical observations, and are laid down with the strictest precision. The planning and executing of maps requires great judgment, when they are so made as to give a just representation of the terrestrial globe, in all its various divisions. The Cosmographic society suppose that the horizontal or stereographic projection is the most eligible, as it bears the greatest resemblance to the globe itself. We are indebted to the celebrated Hubner, formerly rector of the college at Hamburg, for the invention of illuminating maps with different transparent colours, by which the limits of each country are distinguished, after a regular and systematic manner.

XX. We might here add the solution of various problems, explain certain paradoxes, and relate many geographic curiosities; but these particulars would carry us beyond our limits: and beside, they more properly belong to the study of geography itself, and are likewise more curious than useful. The most finished particular map that we know, and which may serve as a patern for all others, is that of Bohemia, by Muller.

CHAP.

CHAP. XVI.

GENEALOGY.

I. GENEALOGY is the science of the origin of illustrious houses, of noble and distinguished families: or an enumeration of the ancestors of any person, together with a summary relation of their several alliances, as well in a direct as collateral line. The term genealogy is derived from the Greek, and is composed of two words which signify the one *Genus* and the other *Sermo:* and from this definition it appears, that this science has two objects, and that a good genealogist ought to know, in the first place, the chronological succession of those sovereign and illustrious houses that are, so to say, at the head of nations; and secondly, he should be able to form, from ancient documents, diplomas, and other authorities, genealogical plans of noble and illustrious families; or tables, in which are inserted, in a regular and uninterrupted series, the generations of such distinguished persons as have descended from those families down to the present day.

II. With regard to the first object, genealogy draws its knowledge from the history of nations themselves: for it is history that furnishes this science with the names of those illustrious personages that have adorned any country or nation; with the dates of their birth, marriage and death, their immediate posterity, their alliances, &c. John Hubner, ancient rector of the college of Hamburg, has published, in four folio volumes, a collection of genealogical tables, wherein he has exhibited, in a regular system, and with admirable order, the genealogy of all the illustrious families, as well ancient as modern, that have existed upon the earth, from the days of the patriarchs down to the present time. It is thus that genealogy restores to history what it has borrowed from it; for it is scarce possible clearly to comprehend the latter, to have a distinct idea of all the revolutions that have occurred among the various nations of the earth, without having tables of this sort before our eyes; without knowing the genealogy of those families that have governed or concurred in the government of each nation.

III. It is not easy to conceive in the construction of such tables, how great a knowledge of history in general is necessary, how many particular histories, memoirs, &c. an author of this sort must read or consult, before he sits down to write, what difficulty he will find in reconciling, with propriety, the frequent contradictions

dictions he will rencounter, in fupplying the vacuities, and in drawing the truth out of an abyfs of darknefs. We cannot fufficiently admire the refolution, affiduity and perfeverance of thofe learned men who have undertaken thofe labours, and have executed them in the greateft degree of perfection of which they are fufceptible. We are obliged to refer our readers to the genealogical tables of M. Hubner themfelves, and to a fhort work which his fon has publifhed, by way of dialogue, to facilitate the underftanding them. Thefe are books that can fcarce be confulted but as dictionaries; and which will be found neceffary, but of which it is impoffible here to make an analyfis, or even to give an abftract. With regard to the learning of genealogy in general, nothing is requifite but fight and memory.

IV. The fecond object of this fcience is the knowledge of the names, the days of the births and marriages, and the alliances of the fovereigns, princes and other illuftrious perfonages, who at this time reign or govern in the world: an object alfo that may have great utility, but in which the underftanding has no fhare. This is the province of the memory alone, and whoever carries in his pocket an almanac, or fhort genealogical dictionary, is as learned on opening his book, as he that has thought fit to load his memory with thefe matters, and which perhaps might

might have been furnished with more important matters.

V. The third and last object of a genealogist by profession, is to elucidate the descent of noble and illustrious families: to enumerate their progenitors, to range them in a regular series, to draw up genealogic plans, to supply deficiencies, to discover affinities from the resemblance of names, and to convert conjectures into demonstrations. It is necessary to make here a few observations. The order of society and welfare of mankind require, that the inhabitants of every country should be ranged in different classes; that there should be different states or conditions in life, and that each state should be honoured according to its rank. The nobles are naturally at the head of all the other states, and on that account ought to be treated with great respect. But for any man to entertain a ridiculous prodigality on account of his origin; to imagin himself formed of different materials from the rest of mankind; to reduce to the mere circumstance of birth all that constitutes distinction among men; to suppose there can be any merit in that which is owing entirely to chance, and cannot have any real effect, and to give to this mere incident, that preference which is due to the talents of the mind and the virtues of the heart, which have real and important consequences: and on this illusive idea, the offspring of vanity and weakness, to imagine himself descended

scended from monarchs, heroes, or even gods, to deduce his race from Jupiter, or to place in his genealogical tables the names of Cæsar, Pompey, Palæologus, Charlemagne, Rollo, Wittekind, &c. these are infatuations that are at once very common and highly ridiculous.

VI. History informs all those who would pique themselves on the antiquity of their race, that the origin of all particular families or houses is lost in the darkness of the middle age; that during the fifth, sixth, seventh and eighth centuries, all Europe was over-run by savage nations, who mixed with the natives of each country: that the Moors and Infidels were a long time in Spain, and the remnants of the Goths, Vandals, Catti, Obotrites and many other like nations in Germany; that in most of the western countries they could neither write nor read, before Charlemagne; that there is not in the whole world any one document relative to any family that lived in the tenth century; that the nobility of Spain and Portugal are naturally descended in part from the Moors and Infidels, and perhaps from the Jews, at least with some mixture of these; that their tournaments and feats of chivalry were the invention of the Moors, as well as their romantic gallantry; that in ancient Germany the nobility were not near so respectable as is commonly imagined; that many of these gentry made a profession of robbing on the high way, and had castles to which they retreated with their booty: that travellers in their litanies begged of God to preserve them from meeting with any of these nobility, and there are still ancient litanies remaining in which their particular names are mentioned; and this practice continued till the fifteenth century; that the magistrates of the cities were then considered as the first rank of the people; and lastly, that no private gentleman must expect to find his name, his origin and family in modern genealogies, and still less in the history of past ages, when writing was so rare, and before printing had facilitated the preservation of such inconsiderable objects.

VII. The laws, the constitutions, and received customs require however, that to be admitted into certain illustrious chapters, or military and other orders, the candidate should be able to prove his quarters; by quarter in heraldry is meant a sheild or scutcheon; sixteen of these are necessary to prove nobility by four descents, in those societies where such sort of nobles only are admitted; this term is derived from an ancient custom of placing on the four corners of a tomb, the scutcheon of the father, mother, grandfather and grandmother of the deceased. There are in Flanders and Germany, tombs that have eight, sixteen, and thirty two quarters. The authenticity of the thirty two quarters is, however, always very difficult to be proved, and frequently liable to much suspicion; the proof
of

of the sixteen quarters is abundantly more easy, as they do not go back to those ages when writing was very uncommon. They may without scruple of conscience assert upon oath, their nobility of four descents by sixteen quarters, as is the custom; whereas, in the proofs by thirty two quarters, it is frequently necessary to admit inscriptions, epitaphs, and other vouchers of a very equivocal nature.

VIII. The nobless form genealogical plans or trees of their families, where the chief, the founder, or the first of whom they have any knowledge, is placed at the bottom, as the trunck from whence all the branches shoot that form the tree; at the extremities of these branches are painted the coats of arms of each ancestor in their natural colours, according to the rules of blazonry; so that the youngest or existing branch of the family is at the top of the tree. We sometimes also see, though but rarely, genealogical columns, the fusts of which are in form of a genealogical tree, whose branches that surround the column bear the arms, cyphers or medals of a family. We think we ought not to say more of so dubious a science, and where there is so little certainty of the truth, that it may be properly called the art of hazardous conjectures.

IX. To conclude, the genealogic systems of sovereign and illustrious houses, and the dignified

fied families of modern Europe, are moving pictures, that births and deaths are inceſſantly changing. The cuſtom of ornamenting our almanacs with theſe, is highly uſeful. We have likewiſe in Germany genealogical tables (eſpecially the manual of M. Schumann, which appears every year at Leipzig) which, being carefully made, furniſh every neceſſary inſtruction relative to theſe matters.

CHAP. XVII.

BLAZONRY.

MANY a ſatiriſt has roundly aſſerted that blazonry and phyſic become ſciences merely by virtue of their terminology; and Deſpreaux ſays,

> Auſſitôt maint eſprit fécond en rêveries
> Inventa le Blaſon avec les Armoiries.

Soon

Soon after, man, fruitful in vanities,
Did blazoning and armory devise.

OLDHAM.

Others, on the contrary, have set too high a value on this art, and pretend to find something marvellous in it. F. Bouhours, the Jesuit, seriously asserts, that the motto to a coat of arms is alone an abridgement of perfection; and Scohier assures us, that the study of blazonry is an abyss of knowledge, and that he who shall apply himself to it for thirty or forty years, will still find that he has some thing to learn. F. Meneftrier, a Jesuit, has not only formed the best treatise that we have on heraldry, but has also given an account of all the writers on this science, as well as on blazonry and genealogy, in different languages; and he makes their number amount to 300. Every author is possessed with a good opinion of the science on which he treats, or else it is likely he would have chosen some other: there are consequently three hundred vouchers that blazonry is an important science. But they who are disinterested and impartial take the mid way between these extremes, and suppose, that if blazonry even does not concur to the emolument of mankind, there are many other sciences that are in the same circumstance, and that it is at least interesting to one order of inhabitants, the nobility; that the establishment of different ranks in society is necessary in a state, and that the knowledge of the
origin

origin and distinguishing marks of the first rank among the people, is not a matter of mere indifference: but at the same time no science should be estimated beyond its real value; and blazonry is certainly inferior to many others, seeing it requires scarce any faculty of the mind, but memory, and is beside loaded with a number of barbarous, and frequently absurd terms.

II. Blazonry, or heraldry, in Latin heraldica, is therefore *the science of distinguishing and decyphering all sorts of arms, and of explai·ing them in their proper and peculiar terms.* The word blazon is derived from the German word blasen, which signifies to sound a horn or trumpet. Tournaments were anciently held in Germany every third year. The nobles or gentlemen who presented themselves at the lists sounded a horn, to give notice of their arrival. The heralds, after examining their claim to the title of gentlemen, sounded their trumpets also to inform the marshals, proclaiming with a loud voice the titles, and describing the arms of those who presented themselves. After any getleman had appeared twice at the tournaments his rank was acknowledged, and they sounded the trumpet only, without making further inquiry. From thence the word blasen was used to signify the practice of examining and describing shields and arms in general; of praising or censuring knights, &c. and the word has since remained attached to the science itself.

III. By

III. By the word arms is therefore meant certain marks of honour expressed by various figures and colours, by which the families of those that bear them are distinguished, or such as appertain to a whole nation, city or province. Thus the several respectable families among the Plebeians and Patricians, cities and provinces, have their peculiar arms; and thus ships hoist their flags with the arms of Hamburg, Bremen, Dantzick, &c. Coats of arms are the same marks of honour accompanied with devices or cyphers, and are peculiar to noble and illustrious families; they are drawn in scutcheons or on banners, and were anciently borne on the shield, cuirass, &c. as they are now on standards, colours, &c. They generally reckon eight different kinds of arms, which are, 1. those of houses or families; 2. those of dignities or employments; 3. those of concession, adoption, or aggregation; 4. those of patronage, as the cardinals take the arms of the popes who have raised them to the purple; 5. those of pretension, or of such countries over which the bearer pretends to have authority; 6. those of fiefs, of domains and substitutions; 7. those of communities, republics, cities, academies, &c. 8. those of succession, which are borne by heirs or legatees. Arms are likewise distinguished into expressive or arbitary. Blazonry is, as we have already said, the method of decyphering and describing these coats of arms.

IV. This

IV. This science begins therefore by investigating the origin of arms, and for this purpose it ascends to the highest antiquity: several curious researches of this nature are to be found in the works of Menestrier and Varennius. There are some learned men who pretend to discover, even in the Old Testament, traces of the first use of arms. They suppose they were first borne on the shoe; and the form of the shield or coat on which the arms are painted, by its resemblance to the leather of a sandal or shoe, they say confirms this opinion. The authors who have wrote on this science have borrowed the assistance of the profane historians of the three ages, and after showing that arms have at all times been used as representations of the dignity of birth, the nobility, alliances, employments, and great atchievements of illustrious men, they bring the history of arms down to the present times, and show what are the coats of arms that are now borne by all the sovereign princes of Europe, and even of all the known world: of illustrious houses, of noble families, of countries, provinces and cities, &c. And to a minute description of these, they add their figures engraved according to the rules of blazonry.

V. To acquire a just knowledge of this art, it is necessary to begin with the study of its terminology, that is, we should learn the terms of blazonry, as well ancient as modern, the number of which is so great, that we might easily
compose

compose of them a considerable vocabulary, or short dictionary; and the more, as it is necessary to add the signification to each of these barbarous terms: for this matter, therefore, we must refer our readers to express treatises on blazonry, as those of Varennius, Menestrier, Andrew Favin, Spelman, Colombiere, Bara, Segoin, Geliot, Philip Moreau, Scohier, and especially to a work intituled, *The Art of Blazonry*, or the *Science of Nobility*, &c. published by Daniel de la Feuille, at Amsterdam, 1695. They will there find the greatest part of the terms of this science clearly explained.

VI. In the next place it is necessary to remark the diversity of colours in the shield, which consist of two metals, four other colours, and two furs. The two metals are Or, and argent, or yellow and white: the four other colours are azure or blue, gules or red, sable or black, and vert or green, called sinople, to which is sometimes added purple or violet. The two furs are ermin and vairy; to which are also added counter-ermin and counter-vairy. They say in the etymology of these denominations, that each of the colours express some celestial or mundane virtue, as, for example, that red is called gules, because all beasts by devouring their prey have the gule or throat boody, or of a red colour; and for this reason gules in blazonry denotes valour, intrepidity, &c. It is evident, however, that the most natural colours are expressed by uncommon

common and fantastic names, merely to render them unintelligible, and by means of quackery to make a science of these matters. These colours are represented in drawings and engravings by points and strokes in different directions, and sometimes crossing each other, as well as by distinct signs and characters. There are still two other colours in blazonry, which are the natural colours of fruit, animals or plants, and that of carnation or flesh colour for several parts of the human body.

VII. The figures that usually compose coats of arms are of three kinds, which are, natural, artificial and heraldic. The first consists of representations of all sort of animals, stars, plants, &c. The second of all that art has produced, and that is of use in life, as habitable buildings, bridges, columns, furniture, dress, instruments, tools, military weapons, &c. The heraldic are all those that fill the scutcheons at equal and alternate distances, of metal and colour, or that have a particular situation allotted to some part of the arms; and are, First, all the divisions of the shield, as parti per cross, per chief, pale, fess, bend dexter, bend sinister, chevron, &c. Second, the chief, the bend, the pale, the bar, the chevron, the cross, the saltier, the orle, &c. Third, the faced, bended, barred, paled, traversed, checkered, lozenged, &c. Fourth, billets, frets, guirons, lozenges, mascles, rustres, &c. It is proper to observe here, that all these terms, this

jargon

jargon of blazonry, was in common use in the eleventh century, when that art began to be in vogue; for then the saltiers, fusils, guirons, rustres, &c. were parts of the armour worn by knights: and we find no author who has mentioned this art before the year 1150.

VIII. With regard to the manner of ranging these figures and colours, the principal rule is, always to put metal upon colour, or colour upon metal; and if any example of the contrary is met with, it is from a particular cause which is to be inquired into. The reason they give for this rule is, that the ancient dress was composed of party-coloured stuffs sewed together, or of cloth of gold or silver; and that they put pieces of gold and silver on the colours, and colours on the gold. Blazonary gives a great number of particular rules for the manner of arranging these figures; for quartering and diminishing arms, &c.

IX. Coats of arms have likewise ornaments that may be called exterior, and are accompanied with marks of honour; such as crowns and coronets, colars of the orders of knighthood, ensigns of employment, supporters, the helmet, crest, and mantle. Crowns and coronets have not been placed on scutcheons till within two hundred years past: they are the distinguishing marks of sovereings and of the nobility; as pope.

emperors, kings, dukes, marquifes, counts or earls, and barons: thefe crowns or coronets are different for each order of fovereigns or noblemen. The arms of a knight are furrounded by the collar of his order; and the marks of the military orders is a crofs with eight points, which is placed behind the fhield, the points only appearing. The marks of dignities and employments are, for example, the tiara or triple crown, with the keys, for the pope; the crofs for a bifhop; the baton for a marfhal; the mace, the mortar-piece, &c. The fupporters are thofe figures which are placed on the fides of the arms of fovereigns and the principal nobility. The helmet is placed over the arms with the crown: the helmet is either open or fhut, or with bars, and is placed in front, or in profile. The creft is an ornament or figure that is placed on the top of the helmet; and in the fame part is likewife fometimes placed a plume of feathers.

X. There is in the laft place, the pavilion, which covers and furrounds the arms of emperors, kings and fovereign princes, who depend on God alone for their inheritance: it is compofed of a chapeau or coronet at the top, and a curtain which forms the mantle. Befides thefe, there is the banner that ferves as a creft; cyphers, mottoes, and feveral other particular ornaments; of which blazonry explains the origin, etymology, diverfity and intention.

XI. We

XI. We shall conclude this article with observing, that the science of blazonry also explains, by its rules and in its peculiar terms, the nature of the banners and colours of sovereigns and states, and especially what relates to the flags of maritime nations. Each nation has its peculiar flag, which is borne by all its vessels, except they be pirates, who make use of all colours to surprise those that are weaker, or to deceive such as are stronger than themselves. The two metals, and all the other colours, are used in the same manner on the flag as on the scutcheon. Blazonry therefore describes the colours and arms that belong to each nation, republic, or maritime city, as well in their armies as in their navies.

CHAP. XVIII.

Of Philology in General.

AMONG all the follies to which mankind are liable, there is no one more futile or more disgustful, than a dispute about words. Just denominations, however, are very necessary if we we would convey clear ideas of what they are intended to exprefs; it is very effential therefore, that the name which is given to each fcience be fuch as precifely expreffes its nature, and gives it thofe characteriftics which diftinguifh it from all others. This maxim does not feem to have been carefully obferved by thofe who comprehend, under the term philology, univerfal literature, who extend it to all fciences, fo that each one may there include whatever he thinks proper; as grammar, rhetoric, poetry, antiquities, hiftory, criticifm, the interpretation of authors, &c. This feems to be not only making ftrange abufe of words, but creating confufion in thofe matters where too much regularity and precifion cannot be obferved. The term philology will not admit of an arbitrary and indeterminate ufe. It is compofed of the Greek words φιλο and λογος, which imply a love or ftudy of languages. It appears therefore, magure all authorities that may be produced, and which in fact form no great argument on this occafion, that philology is nothing more than a general knowledge of languages, of the natural and figurative fignification of their words and phrafes, and, in fhort, of all that relates to expreffion in the different dialects of nations, as well ancient as modern.

II. We fhall not examine here whether Eratoftines, the librarian of Alexandria, who, according to Suetonius, was the firft that was called a philologift or critic, bore that name on account

of

of his being a man of great learning, or because he was highly skilful in languages: or whether, in modern times, Justus Lipsius, Angelus Politianus, Cælius Rhodiginus, Muret and others, have obtained the title of philologists by one or the other of these accomplishments; but as in our system we understand, by the term *erudition*, the universality of the sciences, and by that of *literature*, all which relates to the knowledge of antiquities, so we include, under the term *philology*, a critical knowledge of the languages. This science when justly limited is so extensive, that we are obliged greatly to concenter its objects, in order to give the analysis of it in a succinct from.

III. As we have already treated, in the second book, on grammar, rhetoric, eloquence, poetry and versification, we have there given those general rules which are applicable to all possible languages; and as we shall have occasion hereafter, in the twenty-fourth chapter of this book, to explain the principal precepts of criticism, we shall here confine our observation to the languages themselves, and to those general ideas which philology offers, without leading our readers through all the paths of an immense labyrinth.

IV. Language in general may be divided into,
1. *Ancient languages*; which are those that have become extinct with the people who spoke them, or have been so altered and disfigured, that they

no longer resemble the languages which were spoke by those people.

2. *Oriental languages;* the study of which is necessary in order to the understanding of the text of the holy scriptures, especially the Old Testament.

3. *Learned languages;* which are those that are indispensably necessary in the study of erudition, and particularly literature; which, while there were people in the world who made them their common language, were called living; but as no nation now makes use of them, they are called dead languages, and are therefore to be learned from books or in schools.

4. *Modern languages,* in which are distinguished, first, the common languages of the European nations, and secondly, the languages of the people who inhabit the three other parts of the world.

V. With regard to the languages that were spoken by the first inhabitants of the world, till the destruction of the tower of Babel, there are not now the least traces of them remaining; though some zealous theologians pretend that it was the Hebrew, as it is found in the Bible, or at least the ancient Chaldean; but all this is mere conjecture; and it is certain, on the contrary, that every vestige of those languages has been totally destroyed by time. The ancient languages that that have been in use in the different parts of the world since that period, and the

know-

knowledge of which, more or less imperfect, has come down us, are,

1. The Chaldean.
2. The Syriac and Estrangetic.
3. The Arabic.
4. Coptic or ancient Egyptian*.
5. Ancient Ethiopian.
6. Ancient Indian.
7. Ancient Phænician, which is also called the Ionic Phænician.
8. Punic or Carthaginian.
9. Scythian, and the Scythian of the Huns.
10. Cyrillian.
11. Glagolitic.
12. Braminian or Bracmanian.
13. Æolian or Æolic.
14. Jacobitian.
15. Celtic.
16. Saracen.
17. Ancient Esclavonian.
18. Gothic.
19. Hetruscan.
20. Mangiurian; of which the Maronites, Nestorians, and sometimes the Jacobites made use.
21. Hieroglyphic.
22. Runic.
23. Ancient Vandalian.
24. Ancient Germanic.

* The late M. de la Crofe has made a Grammar and Dictionary of this language, which is in manuscript in the library of the university of Leyden.

25. Gaulic.

25. Gaulic.

And perhaps some others that may be known to philologists. To these may be added,

26. The different alphabets, idioms, and methods of speaking and writing in the middle age.

VI. Philology is therefore employed in making learned researches, not only into these languages, but into many others, which we shall enumerate in the three following chapters. It prescribes rules, lays down precepts, points out principles, furnishes etymologies, and makes all the necessary remarks for the understanding and attainment of every known language. It shows the use that may be made of each particular language; in what country, and by what people, it has been spoken; and explains, as far as is possible, all the obscurities and ambiguities that attend the study of each language.

VII. When the alphabet of a language is once discovered and well understood, we may easily attain, or at least with much less difficulty, the knowledge of the rest. Beside numberless philological works, with which each library is crowded, we have, in Germany, a small treatise that is very curious and very instructive, intitled, The new A. B. C. in a hundred languages: or, fundamental instructions for teaching the youngest scholars not only German, Latin, French

and

and Italian, but alfo the oriental and other languages; as well as the pronunciation and knowledge of thefe different languages : Leipfic, publifhed by Gefner 1743. In this book are contained the alphabets and firft elements of a hundred different languages, as well ancient as modern. This work was reprinted in 1748, and very confiderably augmented, under the title of The mafter of the oriental and occidental languages. To this has been added the Lord's prayer, in two hundred languages, ancient and modern, in the characters proper to each, with the dialect or manner of pronouncing the prayer; which contributes greatly to facilitate the attaining an idea of thefe languages. The author of this equally curious and inftructive book is M. John Frederic Frits; and he was affifted by the Danifh miffionary Schults of Hall. The fucceffors of Homann of Nurenburg have alfo publifhed four geographico-philological maps, defigned by Godfrey Henfel; which bear the following titles : 1. Europa polyglotta, linguarum genealogiam exhibens, una cum litteris, fcribendique modis omnium gentium : 2. Afia : 3. Africa; with the fame title : and 4. America cum fupplementis polyglottis. The four parts of the world are engraved and coloured on thefe maps; but in every country, inftead of the names of its cities and provinces, is feen the beginning of the Lord's prayer, in the characters ufed in that country; fo that with a fingle glance

glance of the eye, we see all the languages that are in use in all parts of the known world. These maps are highly curious, and have doubtless cost the inventors immense labour.

VIII. We have elsewhere remarked, that the books which teach the particular rules of a language are called grammars, rudiments, &c. and those that contain the words and phrases, dictionaries, lexicons, lexical manuals, vocabularies, &c. Philology shews the manner in which these books are to be made, and the precautions that are to be observed to render them instructive and agreeable: the method of treating synonymous terms; the gradations that are among words seemingly synonymous; and many other like matters. It shews also the reciprocal influence which the genius and manners of a people have on their language; and their language on their general method of thinking; their manners, urbanity and refinement.

IX. But as it is impossible to perceive all the force and elegance of the various allusions, metaphors and comparisons in a language, especially in an ancient language, if we are not properly instructed in their manners, customs, ceremonies, laws, arts, sciences and professions, and other peculiarities of the nation by whom they have been used, and whose natural idiom they formed, philology, in order to know the true origin, etymology, and signification of the words,

terms,

terms, and phrases of a language, remounts to the most distant ages, and employing all the aids it can receive from literature, it makes use of antiquities, numismatics, and diplomatics, in fixing the meaning of each term, and mode of expression, and by these means renders languages and authors intelligible, clear and agreeable.

X. Those languages, which are no longer in common use, can only be learned by books or manuscripts. But as these have come down to us by the means of copying, they have consequently been frequently mutilated, altered, diminished and disfigured, by those who have copied them; the text, in general, or at least many passages of these books and manuscripts, is unintelligible at the first reading. From hence there has arose in modern Europe a particular science, that is called the *Criticism of Languages*, which makes a part of philology, and is employed, 1. in examining the authenticity and truth of the text; 2. in discovering and pointing out the means of correcting the text; 3. in restoring such passages as have been altered, omitted, or mutilated; 4. in explaining the true sense of the text; and 5. in establishing a language by these means in its full primitive perfection, and making it perfectly intelligible to modern times. The celebrated M. le Clerc has given us an admirable work on this subject, intitled *Ars Critica*, in which he explains, with equal genius and solidity, the rules of sound philological criticism.

XI. That

XI. That which is of the greatest use in understanding and interpreting an obscure or imperfect passage, or an unintelligible word or phrase, is *confrontation*. The best confrontation is that which is made by comparing an author, book or manuscript with itself; by examining if the same word, matter or phrase, is not repeated elsewhere, or in equivalent expressions. This is the most certain method, and produces an authentic interpretation. The second method is to confront a writer with his cotemporaries of the same nation: and the third consists in comparing him with other authors who have written at different times, but in the same language.

CHAP. XIX.

ORIENTAL LANGUAGES.

THO' most of the languages we have enumerated in the preceding chapter, and many of those we shall mention in the twenty-first, have been, or are still in use in the eastern countries,

countries, we here underſtand, however, by the term oriental, thoſe only which are eſſentially neceſſary to the underſtanding, and interpreting, in an exegetic manner, the holy writings, eſpecially thoſe of the Old Teſtament; and for this reſtriction of the term we have the authority of a great number of learned men, who by the oriental languages underſtand only the Hebrew, Chaldean, Syriac, Arabic and Coptic: to which we ſhall add the Samaritan, Rabbinic and Talmudic. Theſe eight languages merit a more particular examen, as they ſerve to eſtabliſh the foundations of the Chriſtian religion, and make a conſiderable part of the ſtudy of a Theologian.

II. The Hebrew, Arabic and Chaldean, reſpectively claim the right of ſeniority, each of them has its advocates, and the point is not eaſy to be decided. Moſt zealous divines are inclined to favour the Hebrew; and there are ſome of them who pretend that it was the language in which God talked with Adam in Paradiſe, and that the ſaints will make uſe of it in heaven in thoſe praiſes which they will eternally offer to the Almighty. Theſe doctors ſeem to be as certain with regard to what is paſſed as what is to come. Some philologiſts give the priority to the Arabic, and others to the Chaldean. This difference is the more difficult to be reconciled, as Moſes was not born till 2464 years after the creation, and in Egypt; that is to ſay, 700 years after the deſtruction of
the

the tower of Babel, when all languages were mixed and confounded; for we have no proof, nor even any account, that the Hebrew was exempted, and preserved its purity amidst the general confusion. There is not, moreover, at this time, any one work of antiquity existing that is wrote in Hebrew, except the Old Testament, and of that there are even some parts in Chaldaic, and words of that and other languages are to be found dispersed in different parts of it.

III. There is one more remark we must here make. The first time we find the word *Hebrew* in the Bible, is in the 13 verse of the xiv. chapter of Genesis; and it is manifest that Abraham and his descendants took that name from the patriarch *Heber*, the son of Salah, and third grandfather of Abraham: it is therefore evident, that in the time of Abraham this name was that of a family, and not of a people who had a separate language. We are therefore to suppose, that Abraham, and the patriarchs after him, spoke the customary language of the country where they lived; that this language changed by degrees, as all living languages have done and ever will do: that Jacob and his sons having passed into Egypt, they and their descendants, under the name of the Children of Israel, did not preserve the language of their fathers in all its purity; but that they mixed with it many expressions borrowed from foreign languages, and especially

especially from the Egyptian and Coptic: that Moses wrote in the Hebrew language, as the children of Israel then spoke it: that the other books of the Old Testament were wrote still later; and that it is almost impossible for this language to have been preserved without any alteration.

IV. Notwithstanding all this, as the theologians are always very sure of what they say, we shall believe with them that the Hebrew was the first language in the world, and that it was delivered from God himself; for these learned doctors tell us, that the Almighty taught it Adam as soon as he had created him, that he might be able to converse with God; and that he gave him the power of calling all things by their names: in the same manner as in after-time the gift of tongues was communicated to the apostles on the day of pentecost. Albertus in his Hebrew Dictionary finds in each word, in each root, in its letters, and the manner of pronouncing it, the signification of that word. Loescher, in his treatise De causis linguæ Hebreæ, carries this matter still further.

V. Nevertheless, as we have no Hebrew but what is contained in the Holy Bible, this language must naturally be deficient of many words; not only because all the ancient languages, but especially those of the first ages, were not so copious as the modern; but there were in those times fewer objects to be named; and the
sacred

sacred authors moreover had not occasion to treat on all subjects. The Hebrew language however is susceptible of all the ornaments of diction, and is very expressive. It is not, beside, so difficult to learn as some have imagined. The style of the Psalms, of the book of Job, and of all that is wrote in a poetic manner, is the most difficult to understand. That of Isaiah is noble and elegant, worthy of an author who was of the house of David, and the nephew and grandson of a king. But, notwithstanding all the labours of the learned for so many centuries, we are very far from having a perfect knowledge of the Hebrew language: this inconvenience is the greater, as it gives occasion to many imperfect translations, which disfigure the true sense of the original text; and, what is still more, they have founded, on these passages wrong interpreted, a belief of events that have never arrived in the manner predicted; and even sometimes religious dogmas. It is to be wished that a society of men, the most learned in these matters, were formed in order to perfect the knowledge of the oriental languages, and of the Hebrew in particular.

VI. The Hebrew language had originally no vowels. They are marked in the massorets by points under the consonants. This language is wrote and read from the left to the right: it has thirteen letters, which grammarians divide into gutteral, palatic, dental, labial and gingival. They

They now distinguish only five vowels in Hebrew, which are the same as ours, a, e, i, o, u. But they divide each vowel into two or three; as long, short, shortest. The articles, pronouns, &c. are placed after the substantive; and the same word is sometimes substantive, adjective and verb. The punctuation and accent are the objects that require the greatest attention in the Hebrew language; they count near forty accents, and there are many whose use is still unknown; they serve in general to distinguish, 1. the period and its members, as the points do in other languages; 2. to determine the quantity of syllables, and 3. to mark the tone that is to be observed in chanting them. Nineteen of these accents are also called, by grammarians, *distinctivi* or *accentus regii*, and the others *conjunctivi*, *servi* or *ministri*. There is, properly speaking, only one conjugation in this language, which is of itself simple, but is varied in each verb by seven or eight different manners, that form in fact so many different conjugations, and give a great number of expressions, to represent by one word the various modifications of a verb. These are the principal characteristics of the Hebrew, as we find it in the Holy Scriptures; and which, taken all together, forms a very regular and analogous language.

VII. The *Chaldean* is that which was spoken in Chaldea. Some say that it is a dialect derived from the Hebrew, and others that the Hebrew

is

is a dialect of the Chaldean. This language has twenty five letters; the forms of which are very different from the Hebrew.' It is in like manner wrote from the left to the right.

The *Syriac* is also considered as a dialect of the Hebrew. It has twenty two letters, which have the same names with the Hebrew, but are of very different forms.

The *Arabic*, or the language of the Arabians, is in like manner a dialect of the Hebrew. It has twenty eight letters, the names of which have a good deal of resemblance to the Hebrew, but their characters are also very different.

The *Coptic* is the ancient language of the Egyptians, but mixed in process of time with much of the Greek. We have already said, in the preceding chapter, that the late M. de la Crose has in a manner re-established this language, when we scarce knew more than the name of it; and that he has composed a Coptic grammar and dictionary. F. Kircher, it is true, had before published a Coptic vocabulary and kind of grammar, but very incomplete. There are thirty two letters in its alphabet, but the characters are almost entirely Greek. There has been no book found in this language but translations of the Holy Scriptures, or ecclesiastic offices, &c.

VIII. The *Samaritan* is another dialect of the Hebrew. The Samaritans were Jews, and their city Samaria was in Judea. They followed the law of Moses with more rigour, more after the

letter

letter than the Hebrews. There is a Samaritan copy of the Pentateuch, which differs, indeed, but little from that of the Jews in Hebrew, but is wrote in different characters, that are commonly called Samaritan, and which Origen, St. Jerom, and many other writers, as well ancient as modern, suppose to be the first letters of the Hebrews. There are also medals that are called Samaritan; they have Hebrew inscriptions, in characters different from those of our Hebrew bible, and which are called square Hebrew. For a further account of the Samaritan language, consult M. Simon in his customs and ceremonies of the Jews, Eduardi Bernhardi Lexicon Samaritanum, F. Kircher, M. Buxtorff, M. de Spanheim, F. Morin, M. Walton, and a great number of other writers.

IX. The *Rabbinic*, or the Hebrew of the Rabbins, is the language of which they have made use in their works. The body of it is composed of Hebrew and Chaldaic, with divers alterations in the words of those two languages, whose significations they have much extended. They have likewise borrowed greatly from the Arabic. The rest is composed of words taken for the most part from the Greek, with some from the Latin, and others from various modern languages, especially that of the country in which each rabbin lived. For we should remember here, that after the return from the last captivity, they spoke scarce any pure Hebrew at Jerusalem.

rufalem and in Judea, but Greek mixed with fome Hebraic expreffions; the Romans afterward entering Palaftine, and becoming conquerors of that country, fpoke their own language there: and at laft the Jewifh nation was totaly difperfed. We fhall only add, that the Rabbinic is a very copious language, and that there is fcarce any part of fcience of which the Rabbins have not treated, but always with an enthufiafm that is natural to them: there have been among them even poets and orators.

X. The *Talmudic* is another dialect or particular idiom of the Hebrew, in which the Talmud, or the book compofed by the Jews that contains all the explications of their law, is written. This language differs greatly from the pure Hebrew. M. Buxtorff has compofed a Chaldaic, Talmudic and Rabbinic dictionary. We have alfo a work of the emperor Conftantine, intitled Clavis Talmudica; and one of Otto, called Vitæ doctorum Mifnicorum; befide feveral others.

XI. We fhall conclude this article with faying a few words concerning the Hebrew characters in general. Thefe are the ancient Hebrew, the modern Hebrew, the fquare and the Rabbinic Hebrew. The fquare Hebrew is fo called from the form of its letters, which are more regularly fquare, and have their angles better defined than the Hebrew of the Rabbins. The

The faireſt characters in the ſquare Hebrew, are ſuch as reſemble the characters of the Spaniſh manuſcripts: the next are thoſe of the Italian manuſcripts, and then thoſe of the French and German. Many authors ſay that the ſquare Hebrew is not the true ancient character that the Jews wrote from the origin of their language to the captivity of Babylon; but an Aſſyrian or Chaldean character, which they adopted during their captivity and have ſince retained. The Rabbinic is a character not inelegant, and is formed from the ſquare Hebrew by cutting off the greateſt part of its angles.

CHAP. XX.

Of thoſe LANGUAGES that are called dead, and of PALEOGRAPHY.

LANGUAGES in general, properly ſpeaking, form no ſcience that can enrich the mind with real knowledge, but are to be conſidered as introductions to the ſciences; as keys that open to us the sanctuaries of erudition. In order to attain the knowledge of antiquity in its full extent, the knowledge of those languages that were then in use is of great utility: and properly to judge of modern nations, it is almost indispensably necessary to be acquainted with the principal languages which are now used in the world. There are two languages however, which are called *learned* by way of eminence, and those are the Greek and the Latin. The former of these not only enables us to read the masterly productions of genius of ancient Greece, but also to form a true judgment of all its antiquities, and of its different ages, which form the most entertaining and interesting periods for the sciences and polite arts of all ancient times. The latter affords us the means of understanding the original texts of all the admirable works of the most celebrated Latin authors, and of becoming acquainted with the city, republic, and monarchy of Rome, as if we had been present with them: and of forming a solid judgment of those precious Roman antiquities of every kind, that are still remaining among us.

II. But that which has given the Latin an advantage even over the Greek itself, that has rendered it indispensable to every man of letters, and has made it the basis of erudition, is, that during the middle age, and in general in all modern times, the learned of all Europe have made

made it their common and universal language; so that the Latin forms, if we may use the expression, the natural language of the sciences.

III. All that is written in Greek cannot be properly said to be in the same language, for we should carefully distinguish,

(1.) The ancient or literal Greek: an admirable language, in which are wrote the works of Xenophon, Thucidydes, Demosthenes, Plato, Aristotle, Homer, Sophocles, &c. works that have preserved this language in all its purity, and that will make it, with themselves, immortal. There are, however, several idioms, or dialects in this tongue, among which, four are reckoned principal, and these are, 1. the Attic, which is the most esteemed; 2. the Ionic; 3. the Æolic; and 4. the Doric; which was a kind of rustic dialect, and in which are written eclogues, idyls, and other pastorals. We must observe by the way, that all these four dialects are to be found in Homer, and produce an odd effect in an heroic poem, notwithstanding the universal approbation that is given to this poet. The Greek language is very copious in words, and its inflexions are as various as they are simple in most modern languages. It has three numbers, the singular, dual, and plural, and many tenses in its verbs, which afford great variety of expression. The use of the participles of the aorist, and of the preterit, and of compound words, which are very numerous in this language, give it force and brevity without, in the least, diminishing its perspicuity. Proper names have also a meaning in this, as in the Oriental languages, and the learned there find likewise the character of their origin. The dialect itself, or the pronunciation, is sonorous, soft, harmonious and delightful: in a word, the Greek is the language of a polite nation, that had a taste for all the arts and sciences.

IV. (2) *The Greek of the middle age.* The ancient Greek ended at the time that Constantinople became the capital of the Roman empire, though there were after that time several works, and some by the fathers of the church, which were wrote in Greek, and with sufficient purity: but as theology, law, civil and military policy, the alteration of customs and manners, &c. introduced successively a great number of words that were before unknown, these novelties by degrees altered and corrupted the language—— The natural elegance of the ancient Greek was no longer to be found. Those men of exalted genius, who constantly give a true beauty to a language, were no more. And what could be expected from a barbarous age, and from authors that were even below a moderate capacity?

V. (3.) *The modern or vulgar Greek.* It commenced at the taking of Constantinople by the Turks, and is the language that is now commonly spoke in Greece, without any regard to improvement. The wretched state to which the Greeks are reduced by the Turks, renders them indolent, and, by a necessary consequence, ignorant. The policy of the Ottoman Porte does not permit its subjects to apply themselves to study; and that same spirit, which has destroyed the finest monuments of antiquity, which has made, of columns of porphyry and granate, balls for their cannons, has caused the decadency and total destruction of the sciences. The principal difference between the ancient and vulgar Greek consists in the terminations of their nouns, pronouns, verbs, and other parts of speech. There are also, in the modern, many words that are not to be found in the ancient Greek; particles that appear to be expletives, and which custom alone has introduced to distinguish certain tenses of their verbs; names of employments and dignities unknown to the ancient Grecians; and a great number of words taken from modern tongues: which altogether form a spurious language, a kind of jargon. There is a glossary of this language composed by du Cange.

VI. (4.) *The Greek of the New Testament.* The Greek of the Evangelists and Apostles is very different from that of Thucidydes, Xenophon, and Demosthenes. At the time of the birth of our Saviour, Greek was commonly spoke in Judea; for after the last captivity, the people no longer understood Hebrew: their Greek, however, was corrupted, mixed with a great
number

number of Hebraisms; with words and terms that related to the worship, to the laws, policy, manners, and customs of the Jews; by which means it became a vulgar language, a provincial and rude dialect, in comparison of the ancient or literal Greek. He that understands the New Testament will not in consequence understand Homer. It may appear surprising, that Josephus, the Jewish historian, who lived at the time of the destruction of Jerusalem, about forty years after the death of Christ, should be able to write Greek with so much purity and elegance; but he was at once, a courtier, a minister, a general, and a man of letters; had studied the Greek language, and had spoke it at the court of Vespasian in Rome. For the same reason, St. Paul also wrote better Greek than the Evangelists and other Apostles.

VII. From all that has been said, it is apparent how much utility attends the study of the Greek tongue, and how much reason the English have for applying themselves to it from their early youth. There are, moreover, in modern languages, an infinity of terms in the arts and sciences, as most of those in astronomy, mathematics, physic, anatomy, botany, and the names of many machines, instruments, and other modern inventions, that are either altogether Greek, or derived from it, which renders this language in a manner indispensable to a man of real learning. We cannot, lastly, determine, if modern

dern nations pronounce the Greek language in the manner that the ancient inhabitants of Greece did; but it is very probable, that if Demosthenes or Aristides were now to come upon the earth, they would be very far from understanding what our learned men should say to them in Greek.

VIII. The Latin is the second of those languages that are called dead. It was first spoke in Latium, afterwards at Rome, and by means of the Latin church, and of the labours of the learned, has come down to us. The Latin is not an original tongue, but is formed of the Greek, and especially of the Æolian dialect, and of many words taken from the languages of the Osci, the Hetrurians, and several other ancient nations of Italy. It has had different periods of improvement and decadency, which form its different ages.

The first age comprehends the ancient Latin that was spoke in Latium, and cultivated at Rome, from its first foundation, under the reigns of its kings, and in the first ages of its republic. At the beginning, the Latin tongue was, so to say, inclosed within the walls of Rome, for the Romans did not commonly permit the use of it to their neighbours, or the people they conquered: but when they came to perceive how necessary it was for facilitating their commerce, that the Latin tongue should be spoke every where, and that all nations, in subjection to their

their empire, should be united by one common language, they then obliged those they conquered to adopt their language. It is easy to conceive what must have been the original language of a set of freebooters, without manners, and without arts or sciences; this jargon must, beside, have been necessarily mixed with the language of the Sabines, from whom they stole their wives; and with those of several other foreign nations whom they had conquered, or who were incorporated with their republic. But in proportion as the Romans became polished, their language became refined. There are but very few works of the first age now remaining, among which are reckoned those of Ennius, &c.

IX. The second age of the Latin language began about the time of Cæsar, and ended with Tiberius. This is what is called the Augustan age, which was perhaps of all others the most brilliant. A period at which it should seem as if the greatest men, and the immortal authors, had met together upon the earth, in order to write the Latin language in its utmost purity and perfection. This age, and the language of this age, are so well known, and we have so great a number of works produced at this period, as makes it unnecessary for us to say any thing further of it here.

X. The

X. The third age begins with the reign of Tiberius. Seneca seems to have contributed not a little to have deprived the Latin language of its energy and dignity, and to have substituted the little tricks of style in its stead, and sometimes those childish expressions which the Italians call *concetti*. Even Tacitus appears not to have been quite free from these faults; for his concise and sententious style is not that of the golden age; nor likewise is that of the poet Lucian.

XI. The fourth age of the Latin tongue is that of the remainder of the middle age, and the first centuries of modern times, during which, this language fell by degrees into so great a decadency, that it became nothing better than a barbarous jargon. It is to the style of these times that is given the name of low Latin; and, in fact, it was so corrupted, altered, and mixed with foreign expressions, that M. du Cange has formed a voluminous glossary, which contains those words and phrases only that are used in the low Latin, and which we should not be able to understand without such helps. What indeed could be expected from this language, at a time when the barbarians had taken possession of all Europe, but especially of Italy; when the empire of the east was governed by idiots; when there was a total corruption of morals; when the arts and sciences were in a manner annihilated; when the priests and monks were the only men of letters, and were at the same time

time the most ignorant and futile mortals in the world. Under these times of darkness, we must, therefore, rank that Latin, which is called *lingua ecclesiastica*, and which we cannot read without disgust.

XII. The fifth and last age of the Latin tongue is that which began with the sixteenth century, and was that of Leo X, Charles V, Francis I, Henry VIII of England, &c. A happy period, and ever memorable for the restoration of letters, of arts and sciences, of manners, and of the powers of the human mind, which till then seemed to have remained in a perpetual stupor. It is necessary to remember here, that the art of printing was not invented till about the year 1441; and that the manuscripts of the ancient Greek and Latin authors were become extremely scarce and highly valuable; so that but few private persons were able to procure them, and to study the Latin of the Augustan age. But since that time, we have had many Latin works, as well in verse as prose, in a style that we cannot sufficiently admire, and which, though not altogether so pure and elegant as those of the golden age, yet are not much inferior.

XIII. There are, however, in the Latin, and in all dead languages, two great inconveniences which continually attend them, with regard to modern ages. The first consists in the pronunciation.

ciation. As to what concerns the Latin, each nation pronounces it after the manner of its own language, and each of them imagines their pronunciation to be the best. It may be proved, however, by many arguments, that no man now upon earth, pronounces Latin in the same manner as did Horace and Cicero. The second inconvenience is the deficiency of the Latin language with regard to us, as it has not terms whereby to express those inventions and discoveries of every kind that have been made since the existence of the Roman empire. There are no Latin words for any of the furniture that surrounds us, for three fourths of the dishes that come upon our tables, for the dress we wear, for our instruments of war and navigation, for civil and military employments, and, in a word, for all our daily occupations. It is droll enough to hear our authors call a cannon, bombarda; a peruque, capilamentum; and a button of our cloaths, globulus, &c. Whoever shall doubt the propriety of this observation, need only read the essays that some able Latinists have made in our days to write gazettes in that language, and they will there see the pains those writers have taken, and the ill success they have had. We shall say no more of a language which every scholar learns from his infancy, which is taught over all Europe in schools and colleges, and of which there are grammars, dictionaries, and other instructive books without number.

XIV. Paleography

XIV. Paleography is a description of the ancient manner of writing a language from its origin to the most recent time. This denomination is taken from the two Greek words παλαιος *palaios*, and γραφη *graphe*; of which the former signifies, ancient, and the latter, writing. Paleography is not confined to the tracing of the various alterations that have been introduced from age to age in the letters and abbreviations of a language, but it likewise gives an account of the successive changes in the language itself, of the corruptions and barbarisms that have been introduced, or of its improvements, of its acquisitions, and the manner by which it has arrived at the greatest degree of perfection. In a word, it is the history of the revolutions of a language, whether ancient, learned or modern. Abbé Pluche has given, in his Spectacle de la Nature, vol. vii. a paleography of the French language, which may serve as an example, and which we here quote as it is in the hands of every one; who, by consulting it, may easily form an idea of this art.

CHAP.

CHAP. XXI.

Modern Languages.

IF we call all the different dialects of the various nations that now inhabit the known earth, languages, the number is truly great, and vain would be his ambition who should attempt to learn them, though but imperfectly. We will begin with naming the principal of them. There are three which may be called original or mother languages, and which seem to have given birth to all that are now spoke in Europe. These are the *Latin*, *German*, and *Sclavonian*. From the Latin are derived the languages of all those nations which inhabit the southern, and most western countries of this part of the world: From the German, all those of the nations that inhabit the centre and the northern regions: and from the Sclavonian all the languages of the people who dwell in the most eastern part of Europe. The Sclavonian is extended even to Asia; and is spoken from the Adriatic sea to the northern ocean; and almost from the Caspian sea to Saxony. But it must not be imagined from the term Original, which is given to these languages, that they have come down to us

from

from the confusion at Babel without any alteration: No; we have already shown, in the preceding chapter, of what languages the Latin was formed. With regard to the German, it may be very juftly fuppofed to have been the ancient language of the inhabitants of Germany, as the names of their divinities and heroes (Mann, Erta, Hermann, &c. appellative names, which still fignify Man, Earth, Chief of an army) feem to confirm that opinion. But it is indubitable, that the antient German has been mixed and corrupted by the languages of thofe northern nations which in the fourth century deluged Europe: and who, when they penetrated Italy and Africa, did not merely pafs through Germany as an army that marches in regular order, but remained there a confiderable time, and mixed with the natives of the country. All thefe Scythian or Celtic people acquired likewife in Germany the name of Allamands or Germans; fome were called Goths, that is, good; others Quades, or bad; others Huns, or dogs; others Normans, or men from the north; and fo of the reft. And thofe nations were from that time known and diftinguifhed by thefe denominations.

II. With regard to the Sclavonian, it is to be fuppofed that it is in part the antient language of the Celts or Scythians, mixed with fome particular dialects of different eaftern nations. But be that as it may, thefe three languages

guages appear to have produced the following modern tongues:

(1.) From the *Latin* came,
1. The Portuguese.
2. Spanish.
3. French.
4. Italian.

(2.) From the *German*, or *Allamand*,
5. The modern German, which so little resembles the ancient, that it is with difficulty we read the authors of the fourteenth century.
6. The low Saxon or low German.
7. The Dutch.
8. The English, in which almost all the noun substantives are German, and many of the verbs French, Latin, &c. and which is enriched with the spoils of all other languages.
9. The Danish.
10. The Norwegian.
11. Swedish.
12. Dalecarlian.
13. Laplandish.

(3.) From the *Sclavonian*,
14. The Polonese, with a mixture of the ancient Sarmatian.

15. The

15. The Lithuanian.
16. Bohemian.
17. Hungarian.
18. Transylvanian.
19. Moravian.
20. The modern Vandalian, as it is still spoke in Lusatia, Prussian Vandalia, &c.
21. The Croatian.
22. The Russian or Muscovite.
23. The language of the Calmacs and Cossacs.
24. Thirty-two different dialects of nations who inhabit the north-eastern parts of Europe and Asia, and who are descended from the Tartars and Huno-Scythians. There are polyglott tables, which contain not only the alphabets, but also the principal distinct characters of all these languages.

To all these may be added,

25. The modern Greek, or that which is now spoken in Greece.
26. The modern Hebrew, or vulgar language of the Jews, which is also called the German Hebrew, &c. And,
27. The jargon that is called Lingua Franca.

III. The common languages of Asia are,
28. The Turkish and Tartarian, with their different dialects.

29. The

MODERN LANGUAGES.   353

29. The Persian.
30. The Georgian or Iberian.
31. The Colchic or Mingralian.
32. The Albanian or Circassian.
33. The Armenian.

} These languages are spoke by the Greek Christians in Asia, under the patriarch of Constantinople.

34. The language of the Jews in Persia, Media, and Babylon.
35. The modern Indian.
36. The Formosian.
37. The Indostanic.
38. The Malabarian.
39. The Warugian.
40. The Talmulic, or Damulic.

} The Danish missionaries who go to Tranquebar, print books at Hall in these languages.

41. The modern Arabic.
42. The Tangutian.
43. The Mungalic.
44. The language of Balabandu, and the Nigarian, or Akar Nigarian.
45. The Grusinic or Grusinian.
46. The Chinese.
47. The Japonese.

We have enumerated here those Asiatic languages only, of which we have some knowledge in Europe, and even alphabets, grammars, or other books that can give us information concerning them. There are doubtless other tongues and dialects in those vast regions and

Vol. III.   Z   adjacent

adjacent islands, but of these we are not able to give any account.

IV. The principal languages of Africa are,

48. The modern Egyptian.
49. The Fetuitic or the language of the kingdom of Fetu.
50. The Moroccan, and
51. The jargons of those savage nations who inhabit the desart and burning regions. The people on the coast of Barbary speak a kind of Turkish. To these may be added the Chilhic language, otherwise called Tamazeght: the Negritian, and that of Guinea; the Abyssinian, and the language of the Hottentots.

The languages of the American nations are but little known in Europe. Every one of these, though distant but a few days journey from each other, have their particular language or rather jargon. The languages of the Mexicans and Peruvians seem to be the most regular and polished. There is also one called Poconchi or Pocomana, that is used in the bay of Honduras and toward Guntimal, the words and rules of which are most known to us. The languages of North America are in general the Algonhic, Apalachian, Mohogic, Savanahamic, Virginic, and Mexican: and, in South America, the Peruvian, Caraibic, the language of Chili, the Cairic, the Tucumanian, and the languages used in Paraguay, Brasil, and Guiana.

V. We

V. We have already said, that it would be a vain and senseless undertaking for a man of letters to attempt the study of all these languages, and to make his head an universal dictionary of languages; but it would be still more absurd in us to attempt the analysis of them in this place: some general reflections therefore must here suffice. Among the modern languages of Europe, the French seems to merit the greatest attention, as it is elegant and pleasing in itself, as it is become the general language of courts, and even of public transactions, which are now commonly treated in French: with this tongue likewise we may travel from one end of Europe to the other, without scarce having any occasion for an interpreter: and in this language also are to be found excellent works of every kind, both in verse and prose, useful and agreeable. The other nations of Europe, moreover, find great facility in learning it: the protestant refugees of France of both sexes are dispersed over all Europe; the late M. Regnard found some of them even in the mines of Ostrabothnia. The French, moreover, are fond of travelling and of living in foreign countries, and the inhabitants of other nations are desirous of seeing France, which so well deserves to be visited: from whence arises an useful communication between the French and other nations. We have, besides, grammars and dictionaries of this language which give us every information concerning it, and very able masters who teach it: especially such as come from those parts of France where it is spoke correctly; for, with all its advantages, the French language has this inconvenience, that it is pronounced scarce any where purely but at Paris, and on the banks of the Loire. The language of the court, of the great world, and of men of letters, is, moreover, very different from that of the common people: and the French tongue, in general, is subject to great alteration and novelty. What pity it is, that the style of the great Corneille, and that of Moliere, should already begin to be obsolete, and that it will be but a little time before the inimitable chefs d'œuvres of those men of sublime genius will be no longer seen on the stage! The most modern style of the French, moreover, does not seem to be the best. We are inclined to think, that too much concision, the epigrammatic point, the antithesis, the paradox, the sententious expression, &c. diminish its force: and that by becoming more polished and refined, it loses much of its energy.

VI. The German, Italian and English languages, merit likewise a particular application. They have many real and great excellencies, and are not destitute of natural graces. Authors of great ability daily labour in improving them, and what language would not become excellent were men of exalted talents to make constant use of it in their works? If we had in Iroquois, books like those which we have in Italian, English and

and German, should we not be tempted to learn that language? How glad should we be to understand the Spanish tongue, though it were only to read the Araucana of Don Alonzo D'Ercilia, Don Quixotte, some dramatic pieces, and a small number of other Spanish works, in the original: or the poem of Camoens in Portuguese. The other languages of Europe have each their beauties and excellencies. Happy would he be that could know them: but how many other things are there more necessary to be known than languages?

VII. The greatest difficulty in all living languages constantly consists in the pronunciation, which it is scarce possible for any one to attain unless he be born or educated in the country where it is spoke: and this is the only article for which a master is necessary, as it cannot be learned but by teaching, or by conversation: all the rest may be acquired by a good grammar and other books. In all laguages whatever, the poetic style is more difficult than the prosaic: in every language we should endeavour to enrich our memories with great store of words (*copia verborum*) and to have them ready to produce on all occasions: in all languages it is difficult to extend our knowledge so far as to be able to form a critical judgment of them. All living languages are pronounced rapidly, and without dwelling on the long syllables (which the grammarians call *moram*): almost all of them have

have articles which diftinguifh the genders; all the European languages are wrote from the left to the right, and almoft all the Afiatic from the right to the left.

VIII. Thofe languages that are derived from the Latin have this further advantage, that they adopt without reftraint, and without offending the ear, Latin and Greek words and expreffions, and which, by the aid of a new termination, appear to be natives of the language. The privilege is forbid the Germans, who in their beft tranflations dare not ufe any foreign word, unlefs it be fome technical term in cafe of great neceffity. Our moft fcrupulous tranflators would gladly make ufe of the word *menuet*, if they were not fearful of appearing ridiculous.

IX. To conclude; philology is yet deficient of one very important invention; and that is, an univerfal language, or rather an univerfal character, which each nation may read and comprehend in their own language. After like manner, as all European nations underftand the figures and calculations of each other; and as the Chinefe and Japanefe exprefs their thoughts by the fame characters, fo that thefe two nations can read each others writing, though their languages are very different. The late baron Leibnitz was fo far from believing this invention impoffible or impracticable, that he employed himfelf affiduoufly to the ftudy of it: and it is to be imagined that his death deprived Europe of fo important a difcovery.

CHAP.

CHAP. XXII.

DIGRESSION ON EXERCISES.

THE principal intention of this work being to serve as a guide to youth in the carreer of their studies, and especially to give them some salutary advice for the employment of that precious time which they devote to the academy and university, the reader will not be surprised to find, in this and the three following chapters, a very brief analysis of those exercises, arts and sciences, of which a man of letters ought at least to know the names and first principles, though they do not directly appertain to the system of general erudition: of those arts, which may be even called frivolous, but which the wisest legislators have established for the improvement of mankind.

II. How useful, how agreeable so ever study may be to the mind, it is very far from being equally salutary to the body. Every one observes, that the Creator has formed an intimate connexion between the body and the mind; a perpetual action and reaction, by which the body instantly feels the disorders of the mind, and

the

the mind thofe of the body. The delicate springs of our frail machines lofe their activity, and become enervated; and the veffels are choked by obftructions when we totally defift from exercife, and the confequences neceffarily affect the brain: a mere ftudious and fedentary life is therefore equally prejudicial to the body and the mind. The limbs likewife become ftiff; we contract an auwkard, conftrained manner; a certain difguftful air attends all our actions, and we are very near being as difagreeable to ourfelves as to others. An inclination to ftudy is highly commendable; but it ought not however to infpire us with an averfion to fociety. The natural lot of man is to live among his fellows: and whatever may be the condition of our birth, or our fituation in life, there are a thoufand occafions where a man muft naturally defire to render himfelf agreeable; to be active and adroit; to dance with a grace; to command the fiery fteed; to defend himfelf againft a brutal enemy; to preferve his life by dexterity, as by leaping, fwimming, &c. Many rational caufes have therefore given rife to the practice of particular exercifes, and the moft fagacious and benevolent legiflators have inftituted, in their academies and univerfities, proper methods of enabling youth, who devote themfelves to ftudy, to become expert alfo in laudable exercifes.

III. By the term exercifes, we underftand thofe arts in which a man cannot acquire the
leaft

least ability without the dexterity of the body; and consequently they are to be attained by practice only. Such are,

1. Dancing.
2. Riding.
3. Fencing.
4. Leaping.
5. Wrestling.
6. Swimming.
7. Shooting.
8. Games of address, &c. To which may be added,
9. The art of drawing and raising fortifications on the ground; that of turning; of forming and polishing optic glasses, &c.

IV. We shall see, in the twenty-fifth chapter of this book, that many sovereigns have founded particular academies where these exercises are taught, either solely to the young nobility and gentry, or to the citizens in general: or that they have appointed masters in the colleges for the same purpose. These arts cannot certainly be learned without masters; and it is no small advantage to meet with such as proceed on clear and solid principles. Every one who applies himself to the study of the sciences, would do right well to set apart some hours in the day for exercise; and ought not to regard those hours as lost, but as employed in recreations that are even more useful than agreeable.

V. The

V. The last method of expressing our thoughts, the sentiments and passions of the mind by means of the sight, is in the dance; see vol. ii. page 419. Almost from the first accounts we have of mankind by history, we are told of their dancing: we must not imagine, however, that the dances of the first inhabitants of the world, or all those of ancient nations, were like such as are practised in our days: for we cannot suppose that when the king and prophet David danced before the ark, he did it in the step of a minuet or country-dance, as that would present a very strange idea, and not very compatible with our notions of the propriety of manners. We should not have a very high opinion of a king of France or Spain, for example, who should dance before the host in a religious procession, and in the face of all the people. The dance was, among the ancients, sometimes a religious ceremony; and it is said in Ecclesiastes, that *there is a time to dance*. We have already remarked, in the chapter on declamation, that the Greeks used the word *orchesis*, and the Latins that of *saltatio*, in a much more extensive sense than we do that of dancing; and that the theatric declamation, accompanied by artificial gestures, and a determinate expression, was there included; as well as the art of mimics and pantomimes, &c. The translators meeting with the word *orchesis*, ορχησις, and the verb ορχεω, from whence also is derived the term orchestra, and *saltatio*, saltare, or other

other equivalent terms; and finding themselves embarrassed by the indigence of modern languages, and the diversity of our customs, have expressed them by the words dance and dancing; though these convey an idea far more confined. We may, moreover, very easily conceive, that the theatric declamation, as well for the voice as the gesture, might be expressed by notes on a scale, and that after the manner of music they might prescribe the elevation of the voice, or the motion of the hands, by the placing of these notes, and determine their duration by proper marks. Our modern chorographies (of which we shall presently speak) moreover confirm this idea.

VI. But without making further inquiry after matters that are now quite out of use, and consequently objects of mere curiosity, let us examine the nature of modern dancing, by which we understand " the art of expressing the sentiments of the mind, or the passions, by measured steps or bounds that are made in cadence, by regulated motions of the body, and by graceful gestures; all performed to the sound of musical instruments, or of the voice:" and which forms at once an exercise agreeable to the performer, and pleasing to the spectator. For we must not imagine with the vulgar, that dancing consists of a jumble of freaks and gambols. The dances of people of education always express some idea: and it was said of Mlle. Salé,

with

with more propriety, perhaps, than is commonly imagined, *that all her steps were sentiments.* Every minuet forms a kind of pantomime, which describes to the eyes an amorous intrigue. Two lovers salute, they amourously regard each other, they give their hands, they separate, they reproach, renew their love by presenting their open arms, they at last give their hands, and again salute in token of reconciliation. It is the same of all other noble and graceful dances. There is in French a charming little treatise, known by the title of " Characters of the dance and of the lovers," where poetry, music, and the dance, very happily concur to express the various characters and sentiments of those who are under the dominion of love.

VII. Modern dancing is divided into that of the theatre and that of society. Theatric dancing consists, 1. of the performance of a single dancer: 2. of dances by two, three, four, &c. 3. of complete ballets, where the chief dancers sometimes perform alone, and sometimes with the chorus of figure dancers: 4. a dance of two, three, &c. with a pantomime ballet; by which is expressed some fact in real or fabulous history; or some other design, by the dance and by gestures. We have seen *chefs d'œuvres* of this kind in the ballet of Pygmalion, or the animated statue; in the ballet of the Rose; in that of Boreas and Zephyrus, and in many other highly ingenious dances. The invention and com-

composition of these dances belong to the ballet-master, who should constantly consult the poet in his choice of subjects, for the dances of an opera or other dramatic piece. It is insufferable to a man of any taste, to see in the Italian operas, ballets that have scarce ever the least relation to the musical drama: the opera of Titus, for example, is terminated by a Chinese ballet; a very grave and tragic story shall be interlarded with a dance of gay, sportful shepherds. This is to join contrarieties and to produce monsters.

VIII. To express the different characters of the persons who compose a ballet, or any other theatric dance whatever, the subjects they are to represent, and the sentiments they are supposed to entertain, the master of the ballet makes use of the different modes or characters in music, and the steps that are appropriated to each mode; as those of the saraband, courant, louvre, &c. for the grave and serious, and those of the minuet, passepié, chaconne, gavot, rigaudoon, jig, &c. for the gay, lively or comic. All these are comprehended under the name of the high dance, and are always accompanied by a graceful motion of the arms. The art of adapting each of these steps, so as happily to express the various sentiments or emotions of the mind, forms the talent of the ballet-master, and is the greatest merit in the composition of a dance.

IX. For-

IX. Formerly there were scarce any dances exhibited on the theatre but the pavan, of which we shall presently speak, and those that do not rise from the ground in displaying the natural graces, either by the manner of the step or in the attitude: the women especially danced only after this manner; but since M. Durpré, Mlle. Camargo, and their competitors, have shown that the high dance, the noble and graceful, is susceptible of leaps or bounds, and of entrechats or capers of six or eight, the entrechat en tournant, the ail de pigeon, the gargouillade, and many other high steps (which must be seen to be understood) the theatric dance is become more lively and brilliant; and the extraordinary abilities of modern dancers have afforded the masters of the ballet opportunity of greatly varying their subjects, of surprising the spectator to a greater degree, by constantly preserving the graceful in the attitudes, and even in the most difficult steps.

X. They always distinguish, however, in theatric dancing, the high and the low, the noble and graceful, and the serious dance; the high, the grand, and the low comic, the antic dance, the pantomime, &c. Every dancer should apply himself to some particular rank of dancing, and there endeavour to excel, according to the extent of his talents. But there are many who can never rise to any considerable rank in their profession, their utmost abilities only enabling them

them to figure in the chorus, from whence they are called figure dancers. The tumblers and rope dancers are not worthy to be mentioned here, as there is no talent required in their performances, but merely the dint of practice.

XI. With regard to the dance of society, the manner of it is greatly altered in Europe. Formerly, for example, they danced in France and elsewhere the pavan, a grave dance that came from Spain; wherein the dancers made a ring by passing one before the other, like peacocks with their long tails. The noblemen performed this serious dance with a cap of state and a sword; the judges in their long robes, the princes in their mantles, and the ladies with the tails of their robes trailing behind them. This was what they called the grand ball. Such gravity would appear highly comic in our days, as all affectation is now laid aside, and nothing is called serious but what is really so: such mimickries of the majestic, therefore, as these, would be regarded as childish and treated with contempt. In the time of Lewis XIV. they still danced at court and at Paris, amiable vainqueurs, passepiés, sarabands, courants, &c. But all these grand matters have been dismissed, and consigned to the wardrobe of ancient gallantry; from whence, however, they may be one day again brought forth, by inconstancy and by the love of novelty. The modern practice of dancing is confined to the minuet and
contre

contre dances or country dances either French or English. In Germany they still sometimes dance allemandes, suabeans, polonese, &c.

XII. By Choregraphy is meant the art of noting on paper the steps and figures of a dance, by means of certain characters invented for that purpose, which are peculiar to this art and are adopted by most nations. The understanding of these requires an express study. They call the description of a dance, whose steps are expressed with the notes of music, orchesography. Thoinet Arbeau printed, at Langres in 1588, a curious treatise on this matter, which he intitled Orchesography; and he was the first who expressed the steps of the dances of his time by notes, in the same manner that songs and airs are noted. He was followed by the famous Beauchamp. We have several books of English country dances where the choregraphy is placed under the airs. Dancing can be learned only by practice; by the aid of a good master, and by imitating those excellent dancers who are to be met with in the great world. They who would excel in dancing should take particular care in their youth not to contract any bad habits, any steps or attitudes that are awkward, constrained or affected. In the last place, dancing is a matter of agility, an exercise that requires natural talents, which are called forth and cultivated by an able master; and who, at the same time that he teaches his art, enables his
pupils

pupils to deport themselves in society with grace, with ease and dignity.

XIII. Pantomimes are representations of those characters, manners, sentiments, actions and passions of mankind, which may be made the subject of a comedy or other theatric performance; and these representations are exhibited by actors, who express their meanings by looks and imitative gestures, without the aid of words. The word mime is Greek, and signifies an imitator, and the word pan means all or all things; so that the compound term pantomime implies an imitator of all things. This term is now used for the representations themselves; and the performers of these comedies, which are called mimes or pantomimes, have been named mimographists. The ancient historians, rhetors, grammarians and critics, give marvellous accounts of the performances of these mimes and pantomimes. Cassiodorus calls them men whose eloquent hands had, so to say, a tongue at the end of each finger. But when they come to particulars, and give examples of their performances, we see that they were little better than trifles. The following is an instance recorded by Macrobius in his Saturnalia: "Hilas, the scholar and competitor of Pylades, who was the inventor of pantomimes, executed after his manner, before the Roman people, a monologue, which ended with these words, *Agamemnon the great*. Hilas, to express those words, made the gestures of a man who would measure another that was greater than himself. Pylades cries out to him from among the people, *My friend, you justly make your Agamemnon to be a man of great stature, but not a great man."* The people demanded that Pylades should instantly perform the same part; and the people were obeyed. Pylades then represented by his attitude and gestures, the appearance of a man plunged in profound meditation, in order to express the proper character of a great man. As if a man of a moderate or even a low capacity was not sometimes rapt in profound meditation. The people however cried a miracle, and shouted applause. What a pitiful example is this! Not that we imagine another actor could have done better in this instance than Hylas or Pylades, but we think that matters like this, and still less scenes of sentiment, can never be well expressed merely by attitudes or gestures; and that it is a folly to attempt it, or to be pleased with so imperfect an expression.

XIV. The Romans, however, were so charmed with these performances, that the two great pantomime rivals, Pylades and Bathyllus, and their most famous successors, were sometimes well nigh distracting the empire by the parties they occasioned among the people. All these pantomime buffoons were at the same time nothing better than miserable eunuchs, who, to make their performance still more ridiculous, acted

with

with a mask, and consequently could express nothing of that continual alteration which arises in the countenance. In process of time these gestures were accompanied by indecent expressions, witness the mimes of Laberius, which were licentious comedies, and which carried these exhibitions to the heigth of extravagance.

XV. A man of genius in the present age, M. Rich of London, undertook to re-establish these pantomimes of the ancients on his theatre; to supply what was deficient, and to give them the utmost perfection of which they seemed capable. He made choice of happy subjects for these representations; he laid aside, with good reason, the mask; he collected the most able actors; he supported the representation, from the beginning to the end, by an accompanyment of diversified and very expressive instrumental music; to all this he added dances, the striking power of decorations, and the almost miraculous power of machinery. By the assistance of all these resources he has at length made the pantomime an amusing entertainment. He has been since imitated by M. Nicolini an Italian, at Brunswick. We have seen with great pleasure, the birth of Harlequin; Harlequin in the mimes of Hartz; and many other charming pieces of this kind: but as these performances speak more to the senses than to the understanding, we cannot see them very often notwithstanding their charming variety.

XVI. In the laſt place; there are ſometimes dances performed by marionetts, which are puppets that are moved by ſprings, and while they are in motion appear to be animated. Theſe are alſo occaſionally uſed by private and reſpectable companies in the performing of ſome farce, or other dramatic piece. Repreſentations of this ſort are made on a ſmall theatre, agreeable to the ſize of the marionetts. The operator who directs their ſprings is concealed behind the ſcenes, ſo that the wooden actors only appear, and who frequently imitate nature to a remarkable degree. This is an entertainment in fact trivial and imperfect enough, and where a certain perſonage, known by the name of punchinello, is the principal character; and who by his blunders, and ſometimes by his droll ſatires, contributes not a little to diſſipate the ſpectators ſpleen; while the ſublime dramas, eſpecially thoſe of the crying kind, plunge him into more melancholy.

XVII. Though there are in all languages many excellent treatiſes on the art of *horſemanſhip*, as thoſe of the duke of Newcaſtle, baron Hochberg, M. Pluvinel, de la Gueriniere, &c. yet this exerciſe can never be well learned but in the menage or riding ſchool, under the direction of an able maſter, and by riding of managed horſes, as well in their natural as artificial paces. To ſit a horſe gracefully, to make him conform to all our deſires, and to avoid all thoſe accidents,

accidents to which riding is liable, are the three principal points that are proposed by learning this art.

XVIII. The art of *fencing* is likewise to be learned from a master, and by exercising in a school; the master is commonly assisted by a prevot or sub-master. It is under this direction that the scholar learns, by the use of files, the proper manner of holding the sword, and of making the various thrusts, as tierce, quart, second, &c. with rapidity and security; as well as the method of parrying all thrusts that can be made at him. *To give, and not to receive is* the motto of a fencing master. There is, in Italian, a treatise by M. Salvatore, *of the theory and practice of fencing:* and a celebrated work in French, by M. Givald Thibault, intitled *the academy of the sword*; as well as several others that have appeared since.

XIX. *Vaulting* is an exercise by which we learn to perform all feats of the body with ease and address; as leaping into the saddle, or dismounting a horse in a like manner, or ascending some great eminence with dexterity, &c. The masters of this art make use of a wooden horse; of a long sloping table, covered with rushes or such like matter, and of some other machinery, for the convenience of their scholars, and for preventing them from unlucky accidents; which
might

might frequently happen in so dangerous an exercise.

XX. *Wrestling* is an encounter by two men without weapons, in order to try their strength, and to endeavour to throw each other on the ground. This was a famous exercise among the ancients, and we still see the cruel and disgustful remains of it among the English. But this exercise is so violent, so dangerous and repugnant to humanity, that far from exhorting youth to the practice of it, we cannot but endeavour to inspire them with an aversion to it. A wrestler by profession, and a spectator who is pleased with such encounters, are commonly two persons equally despicable.

XXI. The art of *swimming*; or the method of sustaining the body on the water by the motions of the arms and the legs, and by properly holding the breath. This exercise is also very dangerous, but at the same time very healthful, seeing that it unites the advantages of a bath with those of exercise: it is, moreover, very useful as it may sometimes save the life or honour of a man. Pieces of cork or bladders may assist those who are learning to swim, but these are weak securities, and on which, therefore, much dependence ought not to be placed. A boat near at hand, and an able swimmer by his side, afford the learner of this exercise the best securities, and the most confidence where there is a natural timidity. The greatest accomplishment in this art is to be able to dive, and to remain under the water, to fetch matters from the bed of a river or the sea, and to rise again with velocity to the surface of the water. M. Thevenot has published a curious work, intitled the art of swimming, illustrated by figures. Everard Digby, an Englishman, and Nicolas Wireman a Hollander, have also given precepts relative to this art.

XXII. The art of *shooting*, whether with the spring bow, the cross bow, the musquet, or fowling piece, &c. at a mark, at a wooden bird, or in the chase, is likewise not to be neglected. This is an exercise that may be of the greatest utility in life, and depends much on a sharp sight, a steady hand, and on practice, which gives a proficiency in all things.

XXIII. The *games* of *address*, as the dexterity in running at the ring; in the combats of the Spanish bulls; in winter upon the ice with skates; at the mall, tennis, bowls, billiards, and numberless other games that are practised in different parts of Europe, are not so frivolous as they may to some appear. These games constantly afford a salutary exercise to the body, render a man active and adroit, and better disposed for more serious occupations. Great care, however, should be taken by youth not to give themselves up to these, and thereby lose that precious time of which every man of letters ought to be so thrifty and even avaritious.

XXIV. In the last place, the art of drawing and raising fortifications on the ground; that of turning wood, ivory, mother of pearl and even metals; that of polishing glasses, and setting them for optical instruments, &c. all these and many other like matters, belong rather to useful arts than exercises. It is true, a man of sedentary life may apply himself to them by way of relaxing his mind and exercising his body, but these arts are to be learned of those who make them their profession; it is sufficient for us just to mention their names and thereby recal them to the readers memory.

CHAP. XXIII.

DIGRESSION on certain Anomalous Arts and Sciences, or such as do not directly appertain to Erudition.

ACCORDING to the general idea, and the definition we have given of Universal Erudition, at the beginning of this work, the more extensive any man's knowledge is, the more Erudition he may be said to have. We have already remarked, however, that there are several sciences which do not directly appertain to the system of Erudition; and it is of these sciences and arts, that we here propose to say a few words; not so much with a view of making their analysis, and thereby confounding them with those that rightly belong to our system, as to show, that though we have not forgot them, yet we think, that from their nature they ought to be excluded, and not confounded with those that rightly appertain to Erudition, and thereby reduce our system to a chaos. We shall therefore barely mention them, and leave those who may have

particular

particular reasons for thinking them worthy of their study, to apply to some good treatise, or to the practice of them; and this we the rather do, as most of these arts and sciences are not the fruits of genius, but merely employments of the judgment and the memory: are founded on experience, and conducted by the aid of the mathematics, or some other science of which we have already treated, or else are subordinate to, and make a part of politics.

II. (1.) *The conduct of a war* requires the union of the theory of that art, with the practice. Now as that art is included in those which concur in the science of government, we have already mentioned, in the chapter on politics, the illustrious names of those great men who have reduced it to a system, and have laboured in teaching it to the public. It is in these schools that they who are ambitious of shining in the fields of Mars, are to seek for instruction. He only, who joins to a fruitful genius, consummate experience, and a solid theory, deserves the name of a great general.

III. (2.) *The marine*, taking that term in its full extent, and in the manner which a minister of that department, or an admiral, ought to understand it, is a science that comprehends, and supposes a masterly knowledge of many other arts and sciences. It is divided into four general parts, which are, 1. the knowledge of all the

the stores, arms, amunition, and other matters necessary to a ship; and with which the magazines and yards belonging to the admiralty ought to be provided. 2. Naval architecture; which teaches the method of constructing all sorts of vessels or ships. 3. Steerage, or the art of conducting a vessel on the sea. And 4. The art of evolutions, which shews the method of commanding a number of ships together, as squadrons or fleets. We do not know of any complete system, that treats of all these four parts together, but there are a great number that treat of them separately.

IV. (3.) *Commerce*; which comprehends vast knowledge, and forms a science that is very intricate, and highly important. Many celebrated authors have endeavoured to reduce it to a system, and have wrote very instructive treatises relative to it. The grand historical and political treasury of the flourishing commerce of the Dutch is a very curious work; the first chapter contains an interesting history of the commerce of all Europe. The works of M. Savary, especially his great dictionary; the elements of commerce; the political essay on commerce, by the late M. Melon: and many other works which are daily appearing in the commercial states, will greatly facilitate the knowledge of these matters. This science, however, makes no part of Erudition, properly so called.

V. (4.) *Coining*,

V. (4.) *Coining*, or the making of money, requires also various knowledge, the union of which forms a very complex art. The knowledge of all metals, their intrinsic and numerary value, their nature, the degrees of their ductility, the proportion they bear to the exchange, their allay, &c. form the preliminary science of a good master of the mint, who is not so common a character as some may imagine. He must likewise understand the art of founding metals, of forming them into ingots or wedges, of reducing them into planchets, or pieces fit to receive the stamp, and the manner of giving them their proper impression, either by the hammer, or the mill. He should also inspect the refining, assaying, plating, graving of the dyes, &c. There are but few good books on this important subject, or even on the several articles of which it is composed.

VI. (5.) *Mineralogy*, or the art of working mines, whether of metals, stones, fossils, &c. forms also an extensive science, and one that is daily improving by practice, and which practice men of ability now endeavour to reduce into a theoretic system, by those discoveries which they are incessantly making of new principles and new inventions. There have been hitherto but few good books wrote on this subject: however, the directors of mines, and miners themselves, of all the countries of Europe, readily communicate to each other their knowledge

ledge and their discoveries. There is a terminology altogether peculiar to this art, and which, being unintelligible to all but miners, requires a particular study.

VII. (6.) *The venery*, which comprehends not only the art of hunting beasts and fowls, the method of knowing their tracks, and fumets or dung, of defeating their artifices, and of regulating the attendants on the chace, as the huntsmen, hounds, &c. but also the knowledge of woods and forests, of what relates to their growth and preservation; the use of the several kinds of trees they produce, &c. There are numberless authors in all languages, who have wrote on the venery, at the head of whom is the emperor Frederic II. A peculiar terminology forms also an essential article in this art.

VIII. (7.) *Political economy*, as well for the city as the country, has been reduced for some time past, in Germany, into a particular science: a number of authors have wrote large works on it, and, in some universities, professors have been established who make complete courses in it, under the title of *collegium œconomicum, urbanum & rusticum*. It happens, however, unfortunately, that these professors are commonly men who in their studies discuss those matters in a methodical manner, which the husbandman, the shepherd, and the fisherman, learn far better, though more slowly, by a daily practice: the rules

rules these professors give, are, moreover, scarce ever applicable out of their own neighbourhood; for there are not under the sun, any two climates and soils perfectly alike.

IX. (8) Flora and Pomona concur to enrich and decorate our lands, and these goddesses have produced among us the *art of gardening*, which has two parts: the first comprehends the theory and practice of pleasure gardens; and the other regards in like manner, fruit gardens, orchards, kitchen gardens, &c. There are very pleasing treatises on this art, as those of Alexander Blond; M. de la Quintinie; the Solitary Gardener; and many others. The hortulan art was so far improved during the reign of Lewis XIV, and under the direction of M. le Nautre, that we almost despair of ever seeing it carried to a greater degree of perfection. The German gardeners, however, have shown, that in producing forward fruit, they have the priority of all other nations, by the aid of their ingenious hot houses: and England is daily decorated by new pleasure gardens, in a style truly original. The English suppose, that a garden ought to represent a beautiful landscape, formed by nature, and ornamented by art: and not the decorations of a dessert precisely disposed, and cut into spruce figures by the shears. On this principle they form their alleys, basons, slopes, woods, groves, &c. as if nature had produced them; regardless of strict regularity:
and

and this method has a marvellous effect, especially in an extensive plan. The descriptions and plans that have been lately published of Chinese gardens, exhibit also ideas that are new and grand in their kind.

X. (9.) Who could have imagined that the preparation of food for man should have produced so complicated an art as is that of cookry? Thanks to the rapacious appetite and refined taste of the ancient and modern Luculli, we have the celebrated treatise of Apicius, de re culinari, which informs us of the state of cookery among the Romans; and, for that of the moderns, we have Le parfait Cuisinier, Le Cuisinier royal et bourgeois; Le Cuisinier moderne, by M. Chapelle, and a great number of similar works, in almost all languages. But this art and these works belong to the universal erudition of the glutton, the voluptuary, and the parasite, who assert that *a cook is a divine mortal*; and maintain by arguments plausible enough, though falacious, that this art is more useful, and requires more wit and sagacity than metaphysics.

XI. (10.) Let us not here forget to mention an art worthy to be honoured by the whole literary world; an art of all others the most pleasing and most useful: and of which they make a very just eulogy in Germany, by a solemn jublee in honour of its invention: in a word, the *Art of Printing*.

Printing. This art has never been placed on a rank with mechanic professions; and the man of sense still laughs at the superstition and ignorance of those priests who would formerly have made the world believe, that typography was a dangerous art. It would require more than one volume to shew how far this art was known, long since, by the Chinese: in what manner it was invented and improved in Europe by John Faustus of Mentz, John Mentel of Strasburg, Guttemburg, Laurence Coster of Harlem, Nicolas Janson, Aldus Manucius, who invented the Italian characters; Elziver, Blaauw, Westein, and an infinity of able printers of our own days: or if we would describe all the mechanism of this art, the various instruments, materials, and workmen that are employed, and the knowledge and taste that it requires. That relation which we have to letters will not permit us, however, to omit this opportunity of giving a public testimony to the abilities of the celebrated M. Breitkopf of Leipzig, who, after having carried the typographic art to the utmost degree of perfection of which it appears capable, has lately invented the art of printing, by the means of moveable characters or notes, all sorts of music, and that with as much precision as taste and elegance. The mere inspection of this surprising art is sufficient to make every one admire the invention, and be charmed with the execution.

CHAP. XXIV.

DIGRESSION on CHIMERICAL ARTS and SCIENCES.

WHEN meditating on the ambitious views of the human mind, we have frequently said,

> Les écarts de raison, l'ignorance & l' erreur,
> Sont de l'esprit humain l'ordinaire appanage.
> Tout mortel pour monter au rang du Cré·teur,
> Voudroit *savoir* beaucoup, & *pouvoir* davantage.

The deviations from reason, ignorance and error, are the ordinary portion of the human mind. Each mortal, to raise himself to a rank with his Creator, would be able to know much, and to perform more. And in fact, the source of all the chimerical arts, and all the frivolous or pretended sciences, seems to be discovered in these four lines. The desire of being highly learned, or at least of appearing so, has given rise to *the art of divination*, and to all those which are dependant on it. The desire of being powerful and formidable, or at least to appear so, in order to seem to predict, has produced the *magical art*, and all those that attend it. So much for the origin of those matters; we shall now see what history relates concerning them.

II. The ancient inhabitants of Asia, in general, partook of the ardor of their climate, and the Chaldeans, in particular, were the greatest visionaries and the poorest philosophers in the whole world. They saw that there was evil in the world, and they could tell how to ascribe it to the All-perfect Being: for they did not perceive, that the terms, good and bad, convey ideas that are merely relative or comparative, like those of great and little; that there could be no such thing as good, if there were no evil by which it might be compared; and that this proceeds from the very essence of all beings whatsoever. They therefore supposed there were two primordial beings, one of which was the author of all good, and whom they named *Oromasdes,* Divinity or God; and the other the author of all evil, whom they called *Arimanius,* Demon or Devil. They did not perceive that it was a far greater offence to the Divinity to suppose an opposite being, another creator and producer beside him, than to suppose that he had produced an evil that was unavoidable and absolutely necessary, and an evil the idea of which is also constantly relative.

III. When this Arimanius or devil, however, was once invented, they did not fail, according to to the laudable custom of the first ages, and of those warm climates, to give him a figure, and make him serve their purposes. This dogma was not sown in barren land. All priests (except those of the Christian religion) have been at all times ambitious and selfinterested. They have sought after great importance, great authority, and great riches. The belief of a demon became therefore to the Chaldean pagan priests a real treasure; the foundation on which they built their principal authority, and the source from whence they derived their greatest wealth. Without the aid of their demon they would have been overthrown more than once; and for this reason it is that they were constantly so jealous of this dogma, and also drew from it such subtle, lucrative and convenient consequences.

IV. All the east, and afterwards all the west, and in short the whole earth, was soon possessed with this dogma. By constantly pursuing earthly ideas, and human notions, the good being was naturally supposed to reside at one place, and the bad being at another. To the former they therefore assigned a heaven, which they supposed to be over their heads, and gave him a celestial court: to the latter they gave a hell, which they imagined to be under their feet, and assigned him an infernal court. From hence arose their gods and demi-gods, their devils, demons, and spirits of every rank and every kind.

V. But

V. But this was not all. This dogma would have been of little confequence if they had not fuppofed a direct, immediate and particular connexion between the infernal court and mankind who inhabit the earth. Now, as no mortal whatever could perceive this connexion by the aid of his fenfes only, they made of it an occult fcience, which naturally remained in the hands of the priefts and prieſteſſes, the magi, the foothfayers, the augurs, the vifionaries, the priefts of the oracles, the falfe prophets, and other like profeffors, till the time of the coming of Jefus Chrift. The light of the gofpel, it is true, has diffipated much of this darknefs; but it is more difficult, than is commonly conceived, to eradicate from the human mind a deep rooted fuperftition, even though the truth be fet in the ftrongeft light, efpecially when the error has been believed almoft from the orign of the world; fo we ſtill find exifting among us the remains of this Pagan fuperftition, in the following chimeras, which enthufiaftic and defigning men have formed into arts and fciences: though it muft be owned, to the honour of the eighteenth century, that the pure doctrines of Chriftianity, and the fpirit of philofophy, which become, God be praifed, every day more diffufed, equally concur in banifhing thefe vifionary opinions. The vogue for thefe pretended fciences and arts, moreover, is paft, and they can no longer be named without exciting ridicule in all fenfible people. By relating them here, therefore,

CHIMERICAL SCIENCES.

fore, and drawing them from their obscurity, we only mean to show their futilty, and to mark those rocks against which the human mind, without the assistance of a pilot, might easily run.

VI. For the attaining of these supernatural qualifications, there are still existing in the world the remains of,

(1.) *Astrology:* a conjectural science which teaches to judge of the effects and influences of the stars; and to predict future events by the situation of the planets and their different aspects. It is divided into natural astrology, or meteorology, which is confined to the foretelling of natural effects, as the winds, rain, hail and snow, frosts and tempests. In this consists one branch of the art of our almanack makers, and by merely confronting these predictions in the calendar, with the weather each day produces, every man of sense will see what regard is to be paid to this part of astrology. The other part, which is called judicial astrology, is still far more illusive and rash than the former: and having been at first the wonderful art of visionaries, it afterwards became that of impostors; a very common fate with all those chimerical sciences, of which we shall here speak. This art pretends to teach the method of predicting all sorts of events that shall happen upon the earth, as well such as relate to the public, as to private persons; and that by the same inspection of the the stars and planets, and their different constellations.

lations. The cabala signifies in like manner the knowledge of thing that are above the moon, as the celestial bodies and their influences; and in this sense it is the same with judicial astrology, or makes a part of it.

VII. (2.) *Horoscopy*, which may also be considered as a part of astrology, is the art by which they draw a figure, or celestial scheme, containing the twelve houses, wherein they mark the disposition of the heavens at a certain moment; for example, that at which a man is born, in order to fortel his fortune, or the incidents of his life. In a word, it is the disposition of the stars and planets at the moment of any person's birth. But as there cannot be any probable or possible relation between the constellations and the human race, all the principles they lay down, and the prophecies they draw from them, are chimerical, false, absurd, and a criminal imposition on mankind.

VIII. (3.) The frivolous and pernicious art of *Augury* consisted, among the ancient Romans, in observing the flight, the singing and eating of birds, especially such as were held sacred. (4.) The equally deceitful art of *Haruspicy* consisted, on the contrary, in the inspection of the bowels of animals, but principally of victims, and from thence predicting grand incidents relative to the republic, and the good or bad events of its enterprises.

IX. (5.) *Aero-*

IX. (5.) *Aeromancy* was the art of divining by the air. This vain science has also come to us from the Pagans: but is rejected by reason as well as Christianity, as false and absurd. (6.) *Pyromancy* is a divination made by the inspection of a flame, either by observing to which side it turns, or by throwing into it some combustible matter; or a bladder filled with wine, or any thing else from which they imagined they were able to predict. (7.) *Hydromancy* is the supposed art of divining by water. The Persians, accordcording to Varro, invented it; Pythagoras and Numa Pompilius made use of it; and we still admire like wonderful prognosticators. (8.) *Geomancy* was a divination made by observing of cracks or clefts in the earth. It was also performed by points made on paper, or any other substance, at a venture; and they judged of future events from the figures that resulted from thence. This was certainly very ridiculous, but it is nothing less so to pretend to predict future events by the inspection of the grounds of a dish of coffee, or by cards, and many other like matters. Thus have designing men made use of the four elements to deceive their credulous brethren.

X. (9.) *Chiromancy*, in the last place, is the art which teaches to know, by inspecting the hand, not only the inclinations of a man, but his future destiny also. The fools or impostors, who practise this art, pretend that the different parts, or the lines of the hand, have a relation to the internal

ternal parts of the body, as some to the heart, others to the liver, spleen, &c. On this false supposition, and on many others equally extravagant, the principles of chiromancy are founded: and on which, however, several authors, as Robert Flud, an Englishman; Artemidorus; M. de la Chambre; John of Indagina; and many others, have wrote large treatises. *Physiognomy*, or *Physionomancy*, is a science that pretends to teach the nature, the temperament, the understanding, and the inclinations of men, by the inspection of their countenances, and is therefore very little less frivolous than chiromancy; though Aristotle, and a number of learned men after him, have wrote express treatises concerning it.

XI. (10.) In the rank of pretended and dangerous sciences, we may also place those fanatico-mystico-theologic doctrines, which still remain in the world, and those books which spiritual visionaries have wrote on these matters, and which others, equally weak, think they understand. We have had a very renowned genius of this kind, in Germany, named Jacob Bohem, and he has had, for successors, some authors not unworthy of him, and many dark preachers. These are constantly a set of impostors, who cover the truth with impenetrable darkness: who pretend to have some particular lights, secret and occult sciences, on those subjects that
are

are so holy and so important, and which require the utmost perspicuity. A spirit of enthusiasm is always concealed in these doctrines and writings, and it is a spirit that a wise legislator should endeavour to suppress wherever it appears. For, to speak plainly, all mystic theology, except that which is sanctified by the church, is an absurd and frivolous science; seeing it is equally repugnant to the wisdom of God, and to human reason, to say, that the sacred writers, who were inspired by the Holy Spirit, have included in their doctrines, beside the true, rational, clear, and instructive sense, one that is mysterious, hidden, allegoric, and involved, which certain visionaries alone can comprehend, which they alone can discover, and which at the same time is neither instructive nor persuasive: or that a book, dictated by the Supreme Being for the salvation of mankind, should contain enigmas, which a theologian alone has a right to expound.

XII. In order to obtain a great and formidable power, and to be able to produce supernatural effects, mankind have also invented,

(1.) *Magic.* This word was at first taken in a good sense, and signified the art of performing uncommon and marvellous acts, by the aid of certain natural secrets, or at least, such as were so to the vulgar. The magicians of those days were men worthy of esteem, who endeavoured to penetrate the hidden powers of nature by lawful

lawful means. Magic was associated with the mathematics, with physic, and theology. Moses himself, Daniel, Apollonius, Tyaneus, Elymas who opposed St. Paul, the sages of Egypt and Babylon, those of the east that came to seek the king of the Jews, who was just born; and numberless other illustrious personages of antiquity, were all magicians. But in the succession of time, these magi applied themselves to astrology, to divinations, to enchantments, and witchcraft; and by those means became opprobrious, and their science contemptible, its productions being no longer regarded but as illusions, mere jugglers tricks. This art is at present in very little esteem, notwithstanding the distinction that is made between natural and supernatural magic, and all the books which have appeared, and still continue to appear under the former title; which generally contain some pretended secrets, and which would be even trifling and peurile, did experience establish their reality. The authors of these ought at least to publish them under some title less ostentatious, obnoxious, and contemptible.

XIII. (2.) *Necromancy*, or *Negromancy*; an art that would be detestable if it were real, and is ridiculous, because it is chimerical: for by this is meant a communication with demons; the art of raising the dead; and of performing many other miraculous facts by a diabolical power, and by enchantments. This was the pretended art of Merlin and Faustus, and which no longer exists.

exists but on the stage, or in childish romances.

XIV. (3.) *Sorcery*, or *witchcraft*, is the third art that pretends to borrow the aid and ministry of the devil, and to perform miraculous operations by invoking demons, either in obscure retreats, or in the darkness of the night, or in an assembly of wizards or witches, which they call a sabbat. It would require a large volume to relate all the influence which this chimerical and absurd art has had on the minds of weak men in all ages, from the creation of the world to the age which immediately preceded the present: to what degree credulous people have believed it: in how serious and important a manner it has been treated by priests, by princes and magistrates, and what horrible cruelties they have been induced from thence to commit. These magistrates were certainly no conjurers: they no ways resembled them, but in wickedness. Since philosophy has confined sorcery to the wardrobe of ancient reveries; and since wise legislators have prohibited the tribunals from exercising their powers against it, and priests from pretending to exorcisms; there is no longer to be found in the world either demon, sorcerer, witch, conjurer, or sabbat.

XV. (4.) The same severity, however, has not been shown to *Alchymy*, though it justly merits

merits as great a punishment from the prince, and as much contempt from the philosopher. If alchymy were nothing more than the art of dissolving natural bodies, and of reducing them to their original principles; of separating the useful parts of each mixture from the useless, so far from deserving to be decryed, it would be an admirable art; but this is the business of the most exalted chymistry, and we should carefully avoid confounding the arts and sciences. Men have perceived in all ages that by the aid of gold the most difficult matters were to be achieved; that if they possessed the art by which Jupiter rained gold, they should be able to accomplish the greatest enterprises, and that they should not even find any difficulty in obtaining a Danae. A modern Prometheus, however, in order to create gold, does not attempt to steal that celestial fire which is so necessary in the creating of all beings, but contents himself with a fire of coals, which he stirs and blows till all the gold, which the inheritance or industry of his ancestors have given him, passes with the smoak up the chimney. The experience of four or five thousand years has not been sufficient to cure mankind of this frenzy; and what is more surprising, is that those, who are reputed philosophers, countenance this practice, by roundly asserting the possibility of making gold. Now, if the matter were really practicable, a good citizen ought not to assert it, because of the small degree of probability
there

there is of ever discovering the secret, and the certain ruin that would attend a great number of men who should attempt it, and the very trifling advantage the discovery would be to society; for perhaps there is no substance, no metal more useless than gold, considered in its own nature. What a celebrated author, whose memory we otherwise revere, has said in his *letters on the sciences*, with the regard to the philosophers stone in particular, fills us with astonishment. Under a specious appearance nothing is less conclusive than his method of reasoning. For, 1. Wether all matter be homogeneous, or 2. that all the parts of matter are reducible to a certain number of principles, which form the elements of all bodies, or 3. that all the parts of matter are as various in themselves as all the different bodies in nature; which ever of these be the case, it is of no importance with regard to the production and generation of composite bodies; and the consequences which he draws from thence may be equally applicable to the production of plants, animals, &c. The source of this argument proceeds from a certain system in philosophy applied to natural history, in which they suppose that stones and minerals are not produced by a regular generation, common to all other material beings: a system that we find very difficult to comprehend, and concerning which we have elsewhere ventured to propose some doubts. For what we might further say here on the subject of alchymy we refer

refer our readers to the chapter on chymistry in the first book. This science appears to us in so futile a light, that we cannot persuade ourselves to make a more ample analysis of it *.

* As we foresee that what is here said will be liable to much objection, in order to avoid all dispute about words, we entreat these philosophers to resolve the following questions.

1. How can we make gold, any more than silver, copper, iron, lead, stones, fossils, shells, &c.
2. How can we make (produce or create) any substance whatever?
3. Can they conceive that there is in nature two generations, two different manners of engendering; and wherefore?
4. If there be two methods of generation, why should there not be more, 3, 4, 5, 6, &c.
5. Why then do they reject equivocal generation, as a chimera?
6. If there are different generations for stones and metals, may there not be also for insects; and why from saw dust mixed with urine may not fleas be generated?
7. Does it require less effort to create or produce a stone, or a grain of gold, than to make a flea; less art to produce a flea than an elephant?
8. The creating of that which is inanimate, or that which appears to us as such, does it embarrass them less than the production of that which is animated?
9. The great or the little, the immense or the imperceptible, are they not equal in true physics? Is it more difficult to make a rhinoceros than a worm?
10. If they know how to make gold (that is, to create, or at least to change the essence of the elements of matter) they can no longer be surprised at all the miracles which the Egyptian magicians performed before their king Pharoah in the presence of Moses.

(5.) The

XVI. (5.) The panacea or univerſal remedy, the potable gold, and the quinteſſences, are alſo chimeras that uſually accompany the philoſophers ſtone, and of which the diſcovery is equally impoſſible. It is a circumſtance ſufficiently mortifying to the human mind, to ſee ſo many men (not confined in a mad houſe) employ themſelves in ſearch of theſe: to ſee ſo many impoſtors run about the world, aſſuring mankind that they have diſcovered them; and to ſee ſo many weak mortals believe them on their word. From whence can they derive any precepts or rules for ſuch 'inquiries? Who can make the analyſis of arts like theſe? Senſeleſs mortals! you would cure thouſands of diſeaſes by one remedy! you pretend to change the order of nature and the decrees of providence! you would perform a perpetual miracle by prolonging the natural duration of beings and the life of man! And can you think that we will countenance ſuch a chimera?

XVII. Men perceiving that they could no longer impoſe on the credulity of their brethren by magic, ſorcery, necromancy, alchimy and the like, have endeavoured to perſuade them that they could, however, perform great matters by *ſympathy*; and have therefore made of it a myſterious art. That appearance of the marvellous which this pretended ſcience contains, has not failed to give it authority among mankind, and eſpecially among the vulgar. It is

true

true, that we see in nature many effects, the causes of which the most profound and sagacious philosophy has not been able to discover. All these have been ranged under the dominion of sympathy, and the visionaries and mountebanks have assumed full powers, where philosophers have prudently been silent. They have invented sympathetic cures for wounds and other disorders, sympathetic powders, &c. &c. They have deprived both men and horses of all power of motion in the middle of a chace; have caused convulsive or swooning fits, and performed a thousand like matters, at an immense distance. We will here assume an affirmative tone, without fear of being thought presumptuous. Rest assured, reader, that there is no such thing as sympathy, properly so called, and in the manner these quacks understand the term. No one body can ever act upon another, in any manner whatever, at a very great distance, and where all communication is interrupted by the air, or other intervening bodies. It is impossible to reduce into system an art or science, or rather a chimera that is founded on no one principle known to any mortal upon earth. We, therefore, rank what Sir Kenelm Digby, and many others before and after him, have wrote on this subject, with the frivolous and pretended arts.

XVIII. It should seem, that it is on such books as these, which treat on fictitious and dangerous

dangerous arts, that the civil magistrate ought to exercise his authority; on works that serve only to fill the heads of mankind with chimeras, to entice them from their labours or useful studies, and to engage them in ruinous enterprises. Every book that contains reflections which are injurious to the majesty of God; opinions that are inconsistent with the order of society; atrocious libels on government, or calumnies on private characters, are worthy of the flames; or what were still better, of confiscation. There are even some useful and respectable prejudices in the world, which a wise man and a good citizen will never publicly expose; and if any one is rash enough to attempt it, he is worthy of chastisement. But that the magistrates of a nation should be such pitiful reasoners, as to wish to treat a harmless philosopher, who may err in the search of truth, as they formerly treated the poor pretended sorcerers, and as they would have treated Galileo, is the most consummate injustice and absurdity. They seem to say with a loud voice: *Citizens, behold a philosophical work, which is wrote with so much strength of argument, that no one can answer it; but so dangerous that we are afraid you should know the truth, lest it might be prejudicial to you. Here, hangman* (what an expression in the free republic of letters) *do your duty!* The treating of errors in philosophy with too much rigour has impeded the progress of the human mind, more than is easily imagined, by checking the spirit of liberty ever since

since the first invention of printing. There is a religion in the world which produces singular and very fatal effects of this nature. We will by no means name it, but let a book be wrote on any subject whatever, we will engage to tell at any time, whether the author was educated in that religion or not; for there are constantly to be seen some traces of constraint, and of certain prejudices imbibed in early days.

※※※※※※※※※※※※※※※※※※※※※

CHAP. XXV.

DIGRESSION on SCHOOLS, COLLEGES, UNIVERSITIES, and ACADEMIES.

THE man who confines himself to his closet is but rarely visited by the sciences, the arts and belles lettres. To acquire their intimate acquaintance he must seek them in those places where Minerva, Pallas, Apollo and the Muses, have fixed their residence;

dence. Emulation, that strong impulse in the career of all our pursuits, should constantly attend the man of letters from his early youth to the last period of his life; in the school, at college, at the university, in those employments to which his knowledge may lead him, or in those academies of science to which he may be admitted. Emulation is an animating faculty that results from society: and few there are to whom nature has given a genius sufficiently strong to attain an extensive erudition in solitude; who are provided with wings that can bear them, without guides, without models, without companions or supports, to the lofty regions of the empyrean.

II. The most sagacious and most benign legislators have therefore established in their dominions, schools for the arts and sciences, academies, porticoes, Lyceums, another Athens; and judiciously adapting instruction to the age and faculties of mankind, they have founded different institutions for this grand design. But far be from these venerable, these sacred abodes, where the mind is invigorated and enriched, where the heart is purified and formed to benevolence, where social man is prepared for those functions to which he appears to have been destined by his Creator, is enabled to render what nature has made rude and barren, polished, refined, and improved to the greatest degree possible! far from these sanctuaries be all defamers of the sciences! Let them deplore in the midst of desarts, or of uncultivated, savage nations, the crime of having endeavoured, though in vain, to degrade the sciences, the arts, the laws and manners of mankind; let them there lament the misfortune of being possessed with a paradoxical spirit. In giving a general idea of Erudition, we think therefore we should describe the outlines of all those admirable foundations for the cultivation of the sciences, which do so much honour to humanity.

III. *Schools* are either public or private establishments for the instruction of the youngest pupils in the first elements of knowledge; in the rudiments of their native language, and sometimes in Latin; in the first principles of religion, &c. In some schools of Germany the French language is likewise taught. Parents, to be free from the care of their children, frequently send them to school while yet too young. They should remember, that at so early an age the springs of the brain are too delicate to be continually stretched by attention; and all that a child acquires by the faculty of his memory is at the expense of his genius, spirit, judgment, and frequently even of his health. There is at Berlin a grand school which they call *Real*, where, to the languages and the principles of religion, they join instructions for drawing, the first elements of history, of the polite arts, mechanics, and

and of many useful employments. This is a very judicious establishment; and has produced many excellent scholars.

IV. *Colleges* are likewise public institutions for the instruction of youth; and are moreover endowed with certain revenues. They there teach divine and human learning, in halls set apart for that purpose, and in what they call *classes*; where the scholars are raised, acccording to their faculties, and the progress they make, from the lowest class to the highest; which is called *prima*. All civilized nations, from the Jews and Egyptians down to those of the present day, have had their colleges. They there teach not only the languages, but also explain the principal classic authors; the regent of each class pointing out to his pupils, at the same time, their various beauties and defect. The first elements of philosophy, and particularly of logic, are likewise there taught. In a word, youth are there prepared for the university; the foundation of that edifice of erudition, which a still more serious study is to raise, is there laid in their minds: for he who carries nothing with him to the university, will certainly bring no great matters from thence. An establishment of this kind is called in Germany *Gymnasium*, but improperly: for among the Greeks that term was applied to a place set apart for bodily exercises.

V. We

V. We alſo ſee with pleaſure, many countries adorned with academies or colleges, founded by wiſe and generous ſovereigns, for the inſtruction of the young nobility and gentry. In theſe illuſtrious gymnaſiums they are taught not only the ſciences and belles lettres, but ſuch exerciſes alſo as are proper to their birth and rank, and for that ſtation in the world which they are one day intended to fill. Among all the eſtabliſhments of this kind there are in Europe, we know of none that approaches nearer to perfection than the celebrated Carolinum of Brunſwick: the young gentleman there meets, at once, the moſt able profeſſors of the ſciences, the beſt maſters for the languages and exerciſes, and, by the favour and indulgence of a very polite court, the moſt efficacious means of attaining a knowledge of the world; at the ſame time that he acquires every kind of erudition which he may hereafter want. The plan of the Carolinum were well worthy to be here given, as the moſt excellent model, if the bounds of this work would admit of ſuch particulars.

VI. *Univerſities* are foundations that have aroſe from the benevolence, the wiſdom, and policy of the beſt of ſovereigns, for the inſtruction of youth in the higher ſciences. They are formed of communities of the different profeſſors in philoſophy, theology, juriſprudence, and phyſic; who each read lectures in public chairs, on the principles of their ſeveral ſciences, to ſuch

such scholars who attend as their disciples or auditors; and to whom they give, when they have finished their courses, certificates of their qualifications, degrees, diplomas, and the doctorial habit. These professors, moreover, assemble in their respective faculties, to decide such cases as may be presented to them, and come under their proper jurisdiction: and lastly, they assemble in a body, and by uniting the four faculties, they form, under the authority of curators, a chancellor, a rector of each faculty, and with the concurrence of a syndic or secretary, a treasurer, and other subaltern officers, the senate of the university. The first book of this work shows what are the particular sciences that are taught in universities, and come properly under their direction. But modern practice (and a very advantageous practice it is) has introduced at universities, professors of history, of the principal sciences that compose the belles lettres, some of the polite arts, exercises, &c. So that a young man, who devotes himself to study, will find at the university the common source of all the sciences; a source that flows in various streams, and from whence he may at once choose that to which he proposes particularly to apply himself, and at the same time drink as much as he thinks proper of all the rest. This assemblage of all the sciences affords those, who devote three or four years of their life to the acquisition of knowledge, the greatest facility, and the most solid advantages.

VII. The

VII. The university of Paris is, without doubt, the most ancient in Europe. It may be justly dated from the time of Charlemagne. That truly great monarch, after having re-established the eastern empire, endeavoured by every means to enlighten and civilize his people. Alcuinus, Raban, Johannes and Claudius, disciples of the venerable Beda, were called to profess the sciences at Paris. This first establishment was successively improved; and in proportion as the scales fell from the eyes of the people, who were nearly reduced to the state of mere brutes, under the dominion of the barbarians, the youth of every country of Europe repaired to the university of Paris to learn the sciences. As the connexion between nations was not then formed in the manner it now is, as neither posts nor coaches, or other public carriages were yet invented, the university maintained proper messengers, who went once or twice every year into the different countries of Europe, carrying with them letters or messages from the students at Paris, and returning with answers from their relations. The titles of these employments still remain in the university, though their functions have ceased; and many persons of rank now seek and obtain these posts, in order to acquire thereby the right of *committimus*. But since Paris has been crowded with nobility of the first rank, courtiers, soldiers, lawyers, financiers, &c. since it has abounded with public diversions, and with those pleasures and dissipations

tions that are the natural confequences, it is become a refidence too noify, and too feducing for the mufes. Other nations have, moreover, improved on the plan of the univerfity of Paris. Of all the univerfities of Europe, thofe of Oxford and Cambridge in England appear at prefent to approach the neareft to perfection. The great men they produce are a better proof than any other argument. We could wifh always to fee an univerfity a real city of learning; a place confecrated entirely to the mufes and their difciples; that the Greek and Latin languages were there predominant; and that every thing were banifhed from thence which could caufe the leaft diffipation in thofe who devote themfelves to letters.

VIII. We fhall fay nothing here of public libraries, anatomical theatres, printing-houfes, and other like eftablifhments which ought to be found in an univerfity; nor of the regulations and difcipline that are there to be obferved. We have treated on thefe matters in our Political Inftitutes, vol. i. chap. iv. the twelfth and following fections; to which we refer the reader.

IX. *Literary focieties* are affemblies of men drawn together by the love of letters; who are united in the cultivating of fome particular parts of fcience; who make all their feveral labours tend to one determinate point; who are

protected

protected by the state, encouraged, and sometimes rewarded with honours and emoluments by the sovereign. Such are the Royal Society of London; that which is called *Naturæ Curioforum* in Germany; that in the same country for the improvement of the language; and many others. These societies commonly fix their assemblies at some determined place; chuse a president or director, a secretary, &c. but at the same time they admit learned foreigners to be enrolled with them. Before the connexions between the European nations were solidly established, before the invention of posts, gazettes, and literary journals, before navigation was so much improved, and travelling so much practised by learned men, ere yet the art of printing was established, and libraries were formed, in every country, it was permissible to suppose that the muses favoured certain privileged places; and that the arts and sciences were there cultivated with an exclusive advantage. But since these happy alterations have taken place, the learned, the men of genius, the artists of Europe, and of the whole world, form but one republic, in which the inhabitants of the banks of the Tagus, the Seine, and the Neva, have an equal right. Experience shows that men are born every where with the same organs, the same faculties and dispositions of the mind; and that there is no more difference between their mental abilities, than between the oaks of different countries. National distinctions are, therefore, banished

banished from this common republic. Men of great and refined talents are every where scarce. But to attribute to certain climates an exclusive faculty of producing beautiful poems or paintings, is a capricious notion, repugnant to reason, and daily contradicted by experience. Literary societies act very wisely, therefore, in admitting men of ability, of every country, to be associated with them.

X. *Academies*, in the last place, are learned communities, instituted by sovereigns, to improve, encourage, and recompense those who have distinguished themselves in the republic of letters, and excel in the arts and sciences. These establishments are not intended to instruct the ignorant, but to improve the learned, to promote the further advancement of letters, and of the arts; and to reward those who therein excel. To be admitted to the honour of being a member of a renowned academy, is to be crowned with the laurels of Apollo: it is to obtain the blue ribbon in the republic of letters. The royal academy of sciences at Paris, instituted for the cultivation of natural philosophy, mathematics and chymistry: the French academy for promoting the purity of that language: that of medals and inscriptions: the academies *Della Crusca* and *Del Cimento* at Florence: the royal academy of sciences and belles lettres at Berlin, which was projected by the renowned Leibnitz, and founded and perfected by king Frederic;

and

and many others; are immortal inftitutions, highly ufeful in promoting of human knowledge, and infinitely glorious for their founders. To thefe academies alfo foreigners are admitted.

XI. Were it our lot to poffefs powerful authority upon the earth, we would add to thefe brilliant eftablifhments yet one more inftitution; and which, perhaps, would not be the leaft ufeful. We would found an encyclopedic academy for the promotion of univerfal erudition. It fhould be compofed of

  3 Members for theology.
  3 ———— for law.
  3 ———— for phyfic.
  3 ———— for fpeculative philofophy.
  4 ———— for natural philofophy and mathematics.
  4 ———— for eloquence and poetry.
  6 ———— for the polite arts.
10 ———— for hiftory, philology, and literature in general.
  4 ———— fupernumerary members for univerfal erudition in thofe parts where they might be ftill neceffary. Thefe would make in all the number of

40 Academicians. To whom we would add a prefident, and two fecretaries: and we would endeavour to procure the moft able profeffors in every clafs. Thefe illuftrious men, thefe literati

rati of the firſt order, ſhould have before them a ſyſtem of univerſal erudition; like that of which we have traced the outlines in this work. Each of the eight claſſes ſhould labour diſtinctly in thoſe matters that naturally belong to their department; and the produce of their labours ſhould be examined in the general aſſemblies. The deſign of this inſtitution would be to furniſh the world, at the end of a certain number of years, with a complete methodical treatiſe of all the arts and ſciences of every kind of human knowledge. So that each reader would find full information concerning univerſal erudition in general, and every part of it in particular. This work, of more importance than any that has hitherto appeared, might extend to twelve, or perhaps twenty volumes in quarto; and might be enlarged from time to time by ſupplements, containing either new diſcoveries, or eclairciſſements of what had been before given. The public would be thereby enriched with a treaſure that would contain the eſſence of all the knowledge of the human mind. There would be only one book more: but how great would be the value of that book!

CHAP. XXVI.

The HISTORY of the SCICENCES.

HAVING thus finished the analyfis of all the fciences in the concifeft manner we found poffible, it will be neceffary, in order to render the fyftem of univerfal erudition complete, to add a few words here:

1. On the general and particular hiftory of all the fciences, of their origin and progrefs.

2. On thofe authors who have cultivated or enriched the fciences, and who may be called the workmen of erudition. And

3. On the principal means by which the knowledge of thofe authors and their works are to be attained, which are (1) by the criticifms that have been made on them, (2) by the literary journals, and (3) by libraries, as well private as public.

The confideration of thefe objects will be the bufinefs of the three following chapters, and which will finifh this work.

II. Literary

II. Literary hiſtory then informs us of the origin, progreſs, decadence, and re-eſtabliſhment of all the arts and all the ſciences, from the beginning of the world to the preſent day. It is either general, and conſiders erudition in its univerſality; or particular, and treats of each art or ſcience ſeparately.

III. Whenever we ſpeak of mankind, we ſpeak of beings endowed with reaſon, for where ever there are men, there are intellectual faculties. Thus it ever was from the beginning of the world, and thus it will be to the end. The firſt operations of the human mind relate to objects that tend to the preſervation of each individual, and the next are thoſe that ſerve to ſupply his wants. When theſe two objects are gratified, the mind begins to reaſon, it becomes philoſophic without knowing it, and without deſiring it; reaſon and experience endow it, by inſenſible degrees, with knowledge. The firſt men were naturally occupied in defending themſelves againſt the elements, againſt ſavage beaſts, and other men but little leſs ferocious; and in procuring the mere neceſſaries of life. For this reaſon it is, that every ſavage and uncivilized nation, every people who are in continual wars, every people who are in want of thoſe objects that are eſſentially neceſſary for their ſubſiſtence, ever have been, and will be, ſtupid, ignorant, and without arts or ſciences.

IV. The

IV. The first men, of whom we have any account, were born in Asia, on that part of the globe which we call, in our situation, the east. They were, doubtless, born with the same faculties of the mind as all their descendants. When they had obtained security and subsistence, they naturally began to exercise their reasoning faculties. Necessity itself made them soon industrious. We must consequently look for the origin of arts and sciences where the first men dwelt, that is, in the east. History confirms what reason teaches us relative to this matter: it shews what was the state of letters in ancient Arabia, in Egypt, Syria, Babylon, Persia, and among the Phœnicians, the people to whom we owe the invention of writing, and from whom all the arts and sciences seem to have proceeded. It also shews how far the powers of the human mind were extended, in those first ages, by the other nations of the known earth. The monuments that are still remaining of those distant times, as for example, the famous ruins of Palmyra, a city of Syria, near to Arabia the Desart, plainly shew that this first age of the arts and sciences ought not to be forgot or despised; and that the most pleasing inventions are not owing to the Greeks, as the most ancient people excelled in the arts, and it was with much difficulty that the Greeks attained an equal degree of perfection; they could even never give that air of grandeur to their productions, which we discover in the works of their predecessors. It is to be be imagined, moreover, that nations who excelled in architecture, could not be quite ignorant of the other arts and sciences, though the length of time has prevented any monuments of them from coming down to us.

V. There is one material remark we muſt here make: It is aſtoniſhing to ſee, in theſe days, men of the greateſt genius, and otherwiſe of the moſt philoſophic temper, poſſeſſed with the notion of the influence of climates, and aſſign to certain regions, more or leſs torrid or temperate, an excluſive power of invention and execution in the polite arts or belles lettres. A belief in ſpectres, in ſympathies, and a thouſand other chimeras that cannot be ſupported by any argument, is equally rational. Whoever will take the trouble to reflect on what we have ſaid in the third and fourth ſections, can no longer entertain ſo ridiculous an error. We are told that the poetry, and all the other expreſſions of the eaſtern nations, breathe a warmth, a certain fire, an enthuſiaſm that is inimitable by the inhabitants of the cold regions of the weſt. In the firſt place, is there, in fact, any great merit in this enthuſiaſm? Thoſe Hebraiſms, thoſe oriental expreſſions, thoſe extravagant hyperboles, forced compariſons, gigantic images, perpetual fictions, that tumid ſtyle, does it all together produce ſuch amazing beauty? It ſhould ſeem, on the contrary, that the more ſagacity mankind have acquired, the more they have quitted this false sublime, have abandoned the project of continual soaring among the clouds, have been content to remain upon the earth, and there imitate the operations of nature.

VI. The ancient inhabitants of the east, and the Egyptians, were moreover accustomed to express themselves by hieroglyphics, and by all kinds of images. It was a national taste of which their style partook, as well in prose as verse. The psalms of David, and the writings of the prophets are full of these images. It would perhaps be dangerous, and even injudicious to imitate them. Now, if this enthusiasm was the effect of the climate, the modern inhabitants of those countries ought to be possessed of it. But experience proves the reverse. The Orientals of our days are cold and phlegmatic, and have preserved nothing of the ancient warmth, but the fastuous titles of their monarchs. The ancient Greeks were notable babblers, the modern are remarkable for taciturnity. The ancient Romans were grave and thoughtful, warriors, politicians, philosophers; the modern inhabitants of Rome, and of Italy, in general, are lively and splendid, great and florid talkers, but weak in war, subtle, refined, industrious: characters totally opposite. Have these climates changed?

VII. The epochs fatal to the arts and sciences arise from four principal causes. The first is war. A people that are continually in arms, and for ever amidſt the buſtle and din of war, have neither ſufficient opportunity nor reſolution to apply themſelves to the ſtudy and the cultivation of the arts. While Aſia was conſtantly in arms; after Philip, Alexander, and their ſucceſſors, were poſſeſſed with the fancy of being conquerors, when the barbarous and warlike nations entered and eſtabliſhed themſelves in Europe, the muſes, ſtunned by the clamour of war, fell into a profound lethargy. The ſecond cauſe is poverty. A people that are ſurrounded by indigence, are too much occupied with their indiſpenſable wants to buſy themſelves with ſtudy; and if there are any men of uncommon genius, who make the moſt happy advancements, they find in their country neither emulation, encouragement, or reward. In England and Holland, on the contrary, we ſee the arts and ſciences flouriſh under the ſhadow of opulence, and in the midſt of the greateſt commerce. The third cauſe is the abuſe that is made of religion, by debaſing it to ſuperſtition, to fanaticiſm and tyranny; than which nothing is more injurious to the progreſs of the human mind. Thoſe ſhackles, which the clergy ſometimes put on philoſophy, prevent all advancement in learning. The hiſtory of every age and every people ſhew their fatal effects. All is loſt when the church once enjoys this kind of triumph. The annals of the middle age, and of the Grecian empire in the eaſt, ſufficiently prove this aſſertion. The fourth and laſt

last cause is, when a succession of stupid, indolent, ignorant, trifling, and, at the same time, despotic sovereigns, who are enemies to the productions of the mind, reign over a nation for a long time together. The reasons are too obvious, and the examples too odious to be recited here.

VIII. Place, on the contrary, a nation under whatever climate you please; let them enjoy continual peace; introduce wealth and plenty among them; confine the authority of the clergy within due bounds; place on the throne a discerning prince; or give them able and learned ministers and magistrates, and you will soon see arise, as it were from the earth, men of the greatest genius, consummate masters in every art and science. These are the natural causes of the improvement or decadence of the arts: the man of sense will find them without labour, without forming hypotheses, or having recourse to illusions and occult causes, or the different nature of climates. But let us return to our subject.

IX. *The second age,* or bright period of the arts and sciences, was the time that preceded the reign of Philip, that passed under his reign, and during the first years of that of Alexander: a period at which there flourished, in Greece alone, such men of exalted genius as Plato, Aristotle,

totle, Demosthenes, Pericles, Apelles, Phidias, and Praxiteles*.

The third age was that of Cæsar and Augustus, whose memory is rendered immortal by Lucretius, Horace, Virgil, Ovid, Cicero, Livy, Cæsar, Varro, Vitruvius, &c.

The fourth age was that of Charlemagne. This monarch, who re-established the empire of the east, was at once the restorer and father of letters: he was himself as learned as a man could be at that time; he composed several books, and among others a grammar of his own language; he endeavoured to enlighten, not only his natural subjects, but those nations also whom he conquered; he made astronomical observations, and established schools in all his dominions; he enticed learned men into France, and, among others, Alcuinus from England; he reduced the laws and customs of those countries that were subject to his empire into writing: during his repasts he caused the histories of the kings his predecessors, or some of the works of St. Augustine, to be read to him; he drew up the capitularies and ordinances for the church with his own hands; he collected all the ancient verses that related to the renowned actions of the Germans and French, to serve him as memoirs for their history, which he intended to write;

---

* See the introduction to M. Voltaire's Age of Lewis XIV.

he had the holy scriptures translated into the German tongue, &c. It is true that this age favoured somewhat of the barbarous ignorance of the times that immediately preceded, and of the wars by which the reign of Charlemagne was continually agitated: but without the assistance of that great prince, literature had been totally lost: he saved it, collected its shattered remains, did all that it was possible to do at that epoch, and what perhaps no other man would have done in his situation.

X. *The fifth age* was that which is called by the name of Pope Leo X. a period when a private family, that of the Medicis, made prodigious efforts in the re-establishment of the arts and sciences, and which in return concurred in the elevation, in the grandeur and glory of that house. So many learned authors, so many great men have said and wrote that the arts and sciences came from the east, from Greece and Constantinople, to seek an asylum among the western nations, after the taking of that city by the Turks, that it is not without timidity we presume to combat that error. Never was there seen, however, more fanaticism, bigotry, ignorance and stupidity, among any people, than in the eastern empire at the time of the taking of Constantinople. M. Montesquieu says*:

* Causes of the grandeur and decline of the Romans.

" A gross

"A grofs fuperftition, which debafes the human mind as much as religion exalts it, placed all the virtue and confidence of mankind in a ftupid veneration for images; fo that generals were feen to raife a fiege, and lofe a town in order to gain an image."——He continues: "When I think of the profound ignorance into which the clergy plunged the laity, I cannot help comparing them with thofe Scythians, of whom Herodotus fpeaks, who put out the eyes of their flaves, that nothing might divert their attention from their labours." And further on he fays: "The fury of difputation became fo natural to the Greeks, that when Contacuzene took Conftantinople, he found the emperor John, and the emprefs Ann, bufy in a council that was held againft certain enemies of the monks: and when Mahomet the fecond befieged that city, he could not fufpend the theologic animofities; the council of Florence engaging their attention, at that time, more than the army of the Turks."

XI. Now let them fairly tell us, what affiftance could be drawn for the arts and fciences from fuch futile mortals as thefe? What book is there left of all the lower empire that a man of fenfe can bear to read? What monuments of the polite arts are there now remaining, or even what traces of them are to be found in Conftantinople or the eaft? A vaft temple of Sophia, the cathedral of the Greek

empire

empire, a clumfy building, with fo little tafte or knowledge of architecture, as to be a difgrace to the art. No ftatues or bafs-relieves, paintings or fculpture; neither verfe nor profe; in a word, nothing has come to us from the lower empire, that does not prove the decadence and diffolution of the arts and fciences in thofe barbarous and fuperftitious times. How then could they be tranfplanted from thence into Europe? We know very well that certain enthufiaftic Arabs came about that time into Italy, and pretended to great learning; but their writings fufficiently prove their mediocrity. It was not fuch people as thefe that brought the arts and fciences from Afia into Europe, but it was Leo X. Charles V, Francis I. Henry VIII. and the other great princes their cotemporaries, that encouraged and protected them, and had the fatisfaction to fee their benign influence produce men of ability and learning of every kind; fuch artifts as Michael Angelo, Raphael, Titian, Taffo, Ariofto, &c. That in ancient times the arts came from Greece to Rome, we readily believe, becaufe thofe arts were then cultivated with the utmoft fuccefs in Greece: but it is impoffible to draw any thing from a country where it is not to be had. The re-eftablifhment of letters is therefore owing folely to the weftern nations.

XII. *The fixth and laft age* is that which M. de Voltaire calls the age of Lewis XIV. It began

gan about the year 1650, and comes down to the present day. This age is enriched with all the discoveries of those that have preceded it, and has effected more than all the other five put together. The faculties of the human mind have been enlarged to the utmost extent, in every part of Europe, and every civilized nation has made the greatest and most successful efforts, in carrying universal erudition to the highest degree of perfection. It is from the general history of the sciences that we learn all the particular inventions, discoveries and improvements, that have been made in the arts, and in letters during these six ages.

XIII. Independent of these general epochs, literary history likewise informs us of the different revolutions that the arts and sciences have undergone in each particular country. It is here we see the origin, progress, and actual state of letters in Germany, France, Italy, England, Spain, and, in short, in every civilized country of Europe. It is extended, moreover, to the other parts of the world. There are a sufficient number of universal literary histories in all languages, and among others that of professor Stolle of Jena in Germany. These works are very useful, but there are many things in which they are all defective, for they speak more of the authors than of the histories of the arts and sciences themselves. It would require a boundless erudition, the utmost strength of judgment, a refined and

subtle

subtle discernment, an exquisite taste, and an absolute impartiality, to compose such a work of this kind as we have still to wish for.

XIV. In the last place, particular literary history instructs us in the rational history of the several sciences; and this knowledge is indispensable to every man who applies to any particular science with a design to make it his profession. The philosophers ought not, doubtless, to be ignorant of the history of philosophy, or of the different systems that have been invented in all ages: the theologian ought certainly to be acquainted with the various revolutions that have happened in his science; the lawyer would be incessantly liable to error, in the interpretation and application of laws, without a thorough knowledge of the history of jurisprudence: the physician ought likewise to know all the remarkable events that have occurred in his art from the days of Esculapius to the present time; and so of the rest. Whoever shall read with attention this analysis of Universal Erudition, will have an idea sufficiently explicit of those arts and sciences whose history he should endeavour to know. We have, moreover, in our progress marked the principal epochs and revolutions. A work three times as large as this would be scarce sufficient to contain the outlines of the history of all the sciences.

CHAP. XXVII.

Of the Knowledge of AUTHORS, and of BIOGRAPHY.

SOLOMON said, more than a thousand years before the Christian era, *That of making books there is no end.* If we believe the Talmud, the ancient rabbins had innumerable libraries in Arabia. Every one knows that Ptolemy II. king of Egypt, amassed more than two hundred thousand volumes, of which he formed his library at Alexandria; and Demetrius Phalaris, to whom he committed the care of it, promised him to make the number soon amount to five hundred thousand. All these books are lost. There are, however, still remaining in the world so immense a number, that the life of man would be scarce sufficient to read the catalogue: and which would require the lives of many learned men to compose. Whoever has read the work of John Albert Fabricius, doctor in theology, and professor at Hamburg, intitled
Bibliotheca

Bibliotheca Græca, in fourteen quarto volumes, which contains an account of such Greek authors only as have come down to us, and the Bibliotheca Latina of the same author, will be easily convinced on the one hand, that a knowledge of authors (Notitia Auctorum) is indispensable to a man of letters; and on the other, that the study of this part of erudition is so extensive, that a work like this cannot pretend to give any detail of it.

II. We shall endeavour, however, to explain some of its first principles. The knowledge of authors and their works, forms, as we have said, a part of literary history. It is divided into universal and particular, sacred and profane, &c. It distinguishes books and authors,

1. Into those of the ancient, the middle, and modern ages; with regard to the time in which the former have been wrote, and the latter have lived.

2. Into theologic, juridical, medicinal, philosophic, those of literature, philology, &c. according to the matter which each author has treated.

3. Into Hebrew, Chaldaic, Syriac, Arabic, Greek, Latin, German, French, and every other language, ancient or modern, in which any author has wrote.

4. Into prosaic or poetic, according to the nature and species of expression.

5. Into Pagan, Jewish, Mahometan, Christian, &c. according to the religion of each author, and the objects he has embraced.

6. Into

Of Authors.

6. Into sacred, ecclesiastic and profane.
7. Into works that are preserved, and such as are lost.
8. Into authentic writings, and those that are spurious.
9. Into complete works, and such as are mutilated, or fragments.
10. Into books published and unpublished.
11. Into printed books and manuscripts. And
12. Into authors that are called classics, common books, and bibliotheques.

III. With regard to the works themselves, it is necessary (1.) to be well acquainted with their titles, (2.) not to mistake allegorical for natural titles, (3.) when a book has two titles, not to mistake it for two different works, (4.) not to confound two authors that have the same name, as Pliny the naturalist, and the younger Pliny, (5.) to know of how many parts or volumes a work consists, (6.) clearly to understand the titles that are marked by abbreviations, (7.) to be acquainted with all the different editions of a book, and to know which of them is the best, (8.) to know the place, the year and form of each edition, (9.) to know the several editors, (10.) to know if any particular edition be enriched with notes or comments, with a summary, index, preface &c. (11.) if all these are good, indifferent or bad, (12.) to know who is the author of the notes, or if the work have been published *cum notis variorum*, (13.) if

(13.) if the book be divided into chapters or paragraphs, (14.) if the edition be handsomely printed, with a good paper and letter, and be correct, (15.) if a work be ornamented with plates of any kind, (16.) if it has been criticised, and if the critics have attacked the matter, the style, or the author personally, (17.) if the critics have been competent judges or ignorant, if they have been impartial or not, &c.

IV. The title of *classic* is properly given to those Latin books only whose authors lived in the Augustan age, and a little before or after it, that is, at the time the Latin tongue was in its greatest purity, and which began to be corrupted after the reign of Tiberius. These writers being read in the classes at schools, or colleges, are therefore called classic authors; and are regarded as of great authority. It is not, however, very clearly determined what authors ought to be raised to this rank. Aulugelus, in his Attic Nights, makes the classics to be Cicero, Cæsar, Sallust, Virgil, Horace, &c. There is, however, no determinate rule for this matter; but much depends on the order established in each college for the different classes. From the account we have here given of this denomination, it is evident, that there are also Greek authors who merit, and who in fact have the title of classic given them, such as Thucidydes, Xenophon, Demosthenes, Homer, Pindar, &c. For the same reason,

reason, they also call St. Thomas the master of sentences, St. Augustine, &c. the classic authors, whom they quote in the divinity schools; Aristotle in philosophy, and so of the rest. It would be both just and highly useful to make choice, in the principal modern languages, of a certain number of authors whose merit is generally acknowledged; to introduce the reading of them in the classes, and to honour them with the style of classic authors; such for example, in the French language, as abbé Vertot, F. Daniel, Patru, Boileau, Racine, Moliere, Voltaire, &c. The same might be done in all other languages. And since the schools have been purged of the reveries of Aristotle, what prevents our naming Locke, Leibnitz, Newton, and Wolff, as classic authors in philosophy?

V. It is quite necessary to remark here, that the knowledge of those ages and nations of the world which preceded the Greeks, is come down to us only by the informations of the Holy Scriptures, and by the Greek writers. Herodotus is the first historian whose works we have. Of Sanchoniathon, or Sanjuniaton, for example, we have only some fragments recorded by Eusebeus. The works of all those authors likewise, who are said to have lived before Homer, as Orpheus, Musæus, Zoroaster, Linus, Hermes, Trismegistus, Horus, Asclepius, Dares the Phrygian, Dictys the Cretan, Hanno, the books of the Sibyls, and a number of others,

are

are entirely loft: what they now produce as their works, are fpurious pieces, and fabricated very lately. It follows therefore, that all our ancient Erudition can begin only with the Greek authors. Thofe books which lead us to a knowledge of the Greek writers, as well as the Latin, and thofe of modern authors of all nations, relative to the arts, the fciences, and doctrines, are therefore the only guides, the only means we can propofe to thofe who are defirous of applying themfelves to this part of erudition. The reft they muft learn by their daily ftudies; and the only advice we can here give them, is not to be prejudiced in favour of any author, ancient or modern; but to read them with circumfpection, and endeavour to diftinguifh, in the writers of every age, the falfe ftone from the true brilliant.

VI. Among an innumerable number of works that lead to the knowledge of books and their authors, we fhall cite only, 1. Diogenes Laertius, and Eunapius de vitis philofophorum; 2. Gerard Jo. Voffius, de hiftoricis; item de poetis Græcis atque Latinis; 3. Martinus Hanikius, de fcriptoribus rerum Romanarum & Byzantinarum; 4. Bluntii cenfura auctorum; 5. Johannis Alberti Fabricii bibliotheca Græca; 6. ejufd. Bibliotheca Latina; 7. ejufd. Bibliographia Antiquaria; 8. Wolffii bibl. Hæbraica; 9. the bibliotheque hiftorique of M. le Long; 10. the bibliotheque poetique of abbé Goujet. In a word, a word, every art, science, and language, has now its bibliotheque or catalogue of books that treat of such matters as relate to it; and F. Labbe, a Jesuit, has composed a bibliotheque of bibliotheques, which contains merely a catalogue of them, and of the authors of all nations who have made catalogues of books. It is manifest, that a work like this must afford vastly more instruction on this subject, than our limits can possibly allow us to give.

VII. It is not less important to know the character of an author, than to know his works. For this purpose, it is proper to be acquainted with the history of his life; 1. at what time he lived; 2. in what country he lived; 3. his rank by birth; 4. who were his relations; 5. what was his fortune, station, or employment; 6. if he can be suspected of partiality, or is supposed to be disinterested, with regard to the subject on which he treats; 7. what were the principal incidents in his life; 8. what sect or religion he professed; 9. who were his masters, colleagues, or cotemporaries; 10. if he was a married or single man; 11. if he travelled, and many other like particulars.

VIII. To the knowledge of books likewise belongs that of translations: as whether a work be rendered in a faithful, elegant, and agreeable manner or not; into what language each valuable book has been translated; what are the names of the most celebrated translators, as Amiot, Du Ryer, Dacier, &c. in what consists the merit or demerit of each translation, &c. The knowledge of all these matters is only to be acquired by much reading and reflection, and by frequenting the best libraries. By these means also, we are enabled to judge of anonymous works, and sometimes to discover the name of an author who may have thought proper to conceal himself.

IX. Prohibited books are commonly very rare and costly, and at the same time are scarce ever worth the pains of looking after. We do not know three prohibited works that are worth reading: we speak of impious and irrational works, such as the famous book De tribus impostoribus, and the two that resemble it; or of certain fanatical works, which are at constant variance with common sense: or of political treatises that have attacked the government at particular periods, which being past, they have lost all their satire: or of lascivious writings, which are calculated to corrupt the morals of mankind; or such works as fill weak and credulous minds with all sorts of chimeras, as *the Clavicle of Solomon*, &c. All works like these are at best but matters of curiosity, and for the most part excite the readers pity; so that we are tempted to exclaim, *is thunder and lightning necessary to destroy such vermin as these?* It is certain, however, that an exorbitant power in the hands of the

the clergy, and the rigour of the laws in certain countries, have proscribed many excellent works; to which posterity will do justice, and eagerly search after.

X. The knowledge of *manuscripts* likewise appertains to that of authors. The critical art shows the manner of distinguishing their age and authenticity; of reading and explaining them, and the uses to which they may be applied. Morhoff, in his Polyphistor, has an entire and very curious chapter on manuscripts; and C. Arnot has published a discourse De selectis doctorum virorum in manuscripta literaria meritis. The liberality with which the celebrated Magliabechi communicates his own manuscripts, or those of others, and even renders them public, does him much honour, and has gained him great esteem among the learned.

XI. Biography is a title given to those books in general, which contain the life, the history, or actions of illustrious men, who are not sovereign princes; and particularly those of learned men and their works; and sometimes also of saints. This term is composed of two Greek words, the first of which signifies *vita*, and the other *scribo*: this term, however, is but little used by the French writers. The biographies of the most celebrated men of letters are of infinite use in attaining a knowledge of authors: they frequently contain anecdotes that are highly

curious,

curious, and which cannot with any propriety be introduced in a regular hiſtory. There have been many of theſe wrote and publiſhed in England, which are equally replete with entertainment and inſtruction.

XII. How much is it to be wiſhed, that the reading of theſe biographies, theſe lives of illuſtrious literati, might excite men of exalted genius to exert all their powers in the career of ſcience! But how unfortunate if they ſhould there find motives for the contrary? If they ſhould be influenced by the fate of a Tſchirnhaus, who ſpent all his fortune in labouring, with the moſt happy ſucceſs, to enlighten mankind, and to make his name revered by all future ages; who was the glory of his country, and cauſed it to abound with riches. The avarice of moſt bookſellers is the principal cauſe of the great ſcarcity of excellent works: but avarice, ſtill more than other crimes, carries its proper puniſhment with it: the ſlender fortune of moſt authors will not permit them to labour for glory alone, the laurels of Apollo will but badly ſupport a numerous family: from hence proceeds that vaſt number of unfiniſhed works, paid by the ſheet, which fill the bookſellers ſhops, load the ſhelves of each library, and in the end ruin the proprietors. And you, the arbiters of human fate, there are born in your dominions men of rare genius, of unbounded talents: while they live, you allow them a bare ſubſiſtence,

subsistence, or more frequently suffer them to languish in penury, and sometimes die for want. When they are dead you would fain recal them, you would render them immortal by public eulogies and statues. Mighty recompense! Wonderful munificence! But you are your own enemies: you deprive your state of its most valuable subjects, and you deprive yourselves of your brightest glory!

CHAP. XXVIII.

DIGRESSIONS

1. On Criticism;
2. On Literary Journals;
3. On Libraries.

I. NO man has ever yet known the bare titles of all the books that have been written: and no one can ever pretend to have a discernment so strictly just, and a knowledge so universal,

versal, as to be able to form a true, infallible judgment on all subjects, and on every author. It is therefore highly advantageous and necessary that there should be in the world, learned, laborious and judicious men, who should make it their business to point out to the studious part of mankind, such books of each age and nation as deserve to be known; and by a clear, impartial, and skilful examination, to show wherein their merit consists. This sort of learned men are called critics, and their labours, criticisms, or productions of the critical art. This art requires, therefore, both discernment and taste, in order to form a just judgment of the matter, and the style of any work. Such was the science of Scaliger, Erasmus, Gesner, Justus Lipsius, Casaubon, Saumaise, &c.

II. Sometimes by the term criticism is also understood a censure that is made of a work or an author; that malicious trouble which some writers give themselves to find out and publish the defects or inadvertencies of an author. This art is far inferior to the preceding, and in which men of very moderate talents are capable of excelling; by its nature, moreover, it has a strong appearance of a depraved temper. A criticism of this sort, when not strictly just, degenerates into insolence, and becomes at once dull and disgustful; for, as M. de St. Real observes, no critic should be allowed to insult an author for an imaginary or dubious fault.
We

We do not remember ever to have read more than one good criticism of this kind, which is that made by the French academy on the Cid of Corneille, and which for truth and discernment, for that method and politeness which is every where observed, and those interesting and instructive reflections with which it abounds, may justly serve as a model to all others. This is the manner in which those critics, who are desirous to censure, should proceed. But such sort of men have seldom any capacity for just criticism. The occasions are, moreover, very rare, wherein it is allowable to search out, and expose to the public view, the faults of a truly valuable work; and never should critics be permitted to extend their censures to the person of an author, for this is not making instructive criticisms, but rancorous satires, and detestable libels.

III. Let us return to the former rank of sagacious critics. All books are considered as old or new: by the former are meant such works as have appeared before our time; and by the latter those of the present day. A knowledge of the first sort is to be attained from the criticisms that the literati, historians, professors of arts and sciences, have made, and are still making, on them; or from bibliotheques. It is by the literary journals that we are to acquire a knowledge of such works as are daily appearing in the republic of letters.

IV. Most

IV. Most countries of Europe, where the arts are cultivated, abound in these days with literary journals; but these are very far from bearing all those marks of merit which are necessary to render them instructive, entertaining and valuable. These journals are no longer wrote by the ancient authors of the Acta Eruditorum of Leipsig: there is now no Bayle, nor any one like him, concerned in writing them. The modern journalists are commonly men of little ability, who, being unable to produce any work worth printing, let themselves out to some bookseller, and then set up for dictators of Parnassus; summons all new authors to appear before their tribunal, praise or blame, and finally determine their merit, with a matchless effrontery. To what judges are the Montesquieus, Chesterfields, Voltaires, Wolffs, Bernoullis, Eulers, Hallers, and many other truly great men, obliged to submit! M. Voltaire has given, in his miscellanies of literature and philosophy, *Advice to a Journalist*: which they ought every one of them to be able to repeat memoriter. They should well remember, that a literary gazette is like one of politics, in which we look for facts and events that happen daily in the world, and not for the crude remarks of a gazetteer. The public alone has a right to judge of the secret causes of an event, and of the wisdom or folly, the equity or injustice of the actors, as well as of the value of a book, and merit of its author; and does not require to have it pointed out by a journalist.

V. But

V. But the best, and perhaps the only way of acquiring a true knowledge of a book, is to read it ourselves. Books are to a man of letters what tools or instruments are to an artist. What is it that produces so great a degree of perfection in the works of art and industry in England and and France, but the goodness of their tools? What is there that concurs more to the perfection of the works of the mind in all countries, than the abundance of valuable writings? Even the most ingenious poets would produce insipid and trifling verses only, mere trash, if sound learning did not appear in their works, amidst all the brilliancy of expression. The dunce and coxcomb may therefore despise books, but the man of sense is convinced, that there is no important knowledge to be acquired without them: he knows at the same time, however, that every thing in this world has its bounds, and that there are collections of books of necessity, utility and ostentation, and that the latter are ridiculous.

VI. Libraries are either public or private. The former are collected and supported by sovereigns or states. These cannot be too numerous; they form, so to say, the archives of the human mind of all ages; and they should furnish every man of letters with all the instructions for which he may have occasion. They concur very efficaciously in the encouragement and improvement of the arts and sciences in each country: and wherever there is a good public library, the people

people can scarce possibly be totally uncivilized. The muses are fond of those places where they find the most delicious nurture for the mind. We cannot therefore wonder to see in the Vatican, at Versailles, Oxford, and such like cities, the most numerous and excellent libraries that can possibly be formed; and to find that the sovereigns and magistrates permit them to be open to the public at certain seasons, and under the direction of learned and polite librarians, from whom each man of letters may also receive information relative to the authors he should consult on each subject. Nothing does more honour to a prince, or produces more advantage to a state, than establishments of this kind.

VII. With regard to private libraries, every man of sense will consult his own abilities in the extent of his library. We are not to ruin ourselves in the service of the muses. But as the fortunes of men are infinitely various, there is no tracing limits for each individual. Whenever we find a man pretend to learning and be quite destitute of books, we have reason to question his pretensions: and whenever we see a statesman, a general, or financier, who has but very little learning, have a numerous and splendid library, we have good reason to suspect him of ostentation.

VIII. Whoever has read this work with attention, will be able to form a complete system of those subjects which ought naturally to be found

in

in an universal library. The works of the critics, and the bibliotheques, for every art and science, will inform him of the names of all the celebrated authors who have wrote on each subject. Every man of letters has commonly some employment, some station in society, or some kind of study to which he is particularly attached. It is very natural for a principal part of his library to consist of books relative to his profession or his favourite study. Thus a prince's library should contain the best authors on politics; and that of a man of literature of the most celebrated critics. For the rest, those books, which contain instructions for forming a library are so very common that we may safely refer the reader to them; barely adding, that the continual efforts of the learned to enrich the literary world with new productions, causes daily alterations in these plans, so that a bibliotheque, which appeared very complete at the beginning of this century, is very far from being so now. Whoever would collect a judicious and useful library, should certainly consult the best journalists, and endeavour to select such works as appear the most excellent in the republic of letters, and consequently his library will increase as long as he lives.

IX. Thus have we finished our proposed plan; have completed our sketch of Universal Erudition, that is, of all the knowledge the human mind has been hitherto capable of acquiring. When we consider the multiplicity and intricacy

of

of these objects, and when we reflect on the weakness of our own talents, we are still inclined to ask ourselves, if we are really arrived at the end of our labour? There may be still some sciences which we have not mentioned, or at least some nominal science, though it may be already comprised in some other part of Erudition: but we are attached to things and not to denominations, to real objects and not to frivolous distinctions.

X. *Ye studious Youth*, it is to you we consecrate our labours: sometimes peruse this abridgment. You will read a romance, ancient or modern, of a dozen volumes, and many frivolous and voluminous works. Why therefore can you not read three volumes? But if you would attempt thoroughly to understand all the arts and sciences we have here indicated, know, that neither the life of man, nor the limits of the human understanding, are sufficient for such a project. If you read this work, however, as you read a romance, you will receive but little advantage: but if you shall seriously study it; if by means of it you acquire a just idea of Universal Erudition, and if from amidst this mass of sciences you shall make a judicious choice of those to which you will particularly apply yourselves, you may become truly learned; and perhaps you will owe us some obligation to your latest hour.

F I N I S.

www.ingramcontent.com/pod-product-compliance
Lightning Source LLC
Chambersburg PA
CBHW022141300426
44115CB00006B/297